CREATIVE CANDLES

CREATIVE CANDLES

OVER 40 INSPIRING PROJECTS
FOR MAKING AND DECORATING
CANDLES FOR EVERY OCCASION

CHANCELLOR
PRESS

CONTENTS

INTRODUCTION

*T*HE SOFT GLOW OF CANDLELIGHT HAS GROWN INCREASINGLY POPULAR. COMPARED TO THE HARSHNESS OF AN ELECTRIC LIGHT BULB THE CANDLE FLAME IS ROMANTIC AND FLATTERING. CANDLELIGHT CAN TRANSFORM EVEN THE MOST DREARY SURROUNDINGS INTO A MAGICAL WORLD. THE VERY SIMPLEST PAIR OF CANDLES CAN ADD ATMOSPHERE TO THE SUPPER TABLE.

Since the early 1960s it has been recognized that candlemaking is a creative and inspirational craft. The basic techniques are easy to master, you need no special skills, and learning the basics is no more difficult than learning how to make an omelette. Once you have mastered the first steps it is great fun to explore more sophisticated ideas, This book describes some of them – but enormous enjoyment is to be had from thinking out new techniques and ideas of your own.

Wax is the most beautiful material to work with. You can dye it, mold it shape it, and scent it. It is malleable when soft – few people can resist the fun of playing with molten wax – and permanent when it has been dyed, molded and hardened.

When you make candles you can also control the way they burn. Candleburning has become almost an artform in itself. A large glass bowl filled with floating candles and colored glass beads looks spectacular, and the effect of the candlelight reflected in the beads and water is dramatic.

Candles make wonderful gifts – especially when they are personalized with decorations. You can make candles inscribed with friends' names, birthday date, age, or astrology sign. It is easy to transform a basic candle for a special occasion such as Christmas or Valentine's day. You can make your most precious creations refillable. It is also fun to make your own holders and think of new ways of displaying your creations so they will look most effective. There are several ways in which you can brighten up candles you have bought ready made – many of them are described in this book. Once you become proficient you can even sell your candles on craft stalls or to stores – most candle factories started with small beginnings!

BASIC TECHNIQUES

EQUIPMENT AND MATERIALS

*Y*OU CAN MAKE CANDLES WITHOUT ANY SPECIALIST EQUIPMENT. MANY ORDINARY HOUSEHOLD OBJECTS MAKE EXCELLENT MOLDS, AND YOU CAN USE YOUR EXISTING SAUCEPANS TO HEAT THE WAX. HOWEVER, ONCE YOU HAVE PASSED THE EXPERIMENTAL STAGE YOU WILL PROBABLY WANT TO BUY SOME CUSTOM-MADE EQUIPMENT, PARTICULARLY A VARIETY OF DIFFERENT TYPES OF MOLDS.

CONTAINERS

Most candle making involves heating wax. It's important that the wax does not overheat, otherwise it ignites. Therefore it's also important to use the right type of container for heating it.

DOUBLE CONTAINERS

Heating wax should always be done in a double container with water in the bottom half. The reason for this is that wax heated in this way cannot become too hot. Wax heated to beyond 212°F (100°C) will ignite. You can improvize, using an ordinary saucepan with a metal bowl or foil container fitted over it.

METAL CONTAINER FOR MELTING WAX

POURING JUG   FOIL CONTAINER

FOIL CONTAINERS

For smaller quantities of wax, kitchen foil containers are excellent. You can place a container directly into a saucepan filled with water, but make sure it does not touch the bottom of the pan or the wax will overheat. The container will float on the water if you do not overfill it. If you want to fill it right to the top, place it on a trivet in the bottom of the pan.

Alternatively, if you have a large saucepan and a large enough trivet, you can heat up several small foil containers of wax at the same time.

This can be useful when you are making candles which need several colors, for example, layered candles.

You will often find that you have small quantities of colored wax left over, and you may wish to keep a color for a future project. Again, foil containers are ideal for this purpose.

POURING JUG

Use a metal jug with a lip to pour wax into the molds.

DIPPING CANS

Many candles need to be dipped in a deep container of wax. You can buy deep dipping cans from specialty shops or you can improvize, using old oil cans or catering-size food tins. Whatever you use, never heat the can directly over your heat source—always place it in a saucepan of water.

MOLDS

There are any number of ways of molding candles. You can improvize your own molds from everyday objects, make your own from latex, or buy ready-made molds.

IMPROVIZED MOLDS

Many ordinary household objects make excellent molds. Small cake tins, for example, make wonderful molds for floating candles. Cardboard milk cartons, yoghurt pots, dishwashing liquid containers, cardboard tubes from inside paper towels or toilet paper— all these and many other everyday objects provide suitable molds for the keen candle-maker. Even cut-off sections of drainpipe can be used for this purpose.

MAKING MOLDS

Roll up pieces of cardboard and use them as an open-ended mold or, when you are more experienced you may want to think about making your own mold out of latex.

READY-MADE MOLDS

Many craft shops and specialist candle-making equipment companies make wonderful candle molds in both basic and novelty shapes.

OTHER USEFUL EQUIPMENT

Apart from a very few other specialized items, most of the equipment you need can be found in your own kitchen.

CUTTING TOOLS

A sharp craft knife and stencil cutters are very useful tools for many of the techniques illustrated in this book.

THERMOMETERS

A thermometer is not essential, but it does help. There are occasions when you will want to use wax heated to more than 180°F (80°C), and you can do this safely using a thermometer. You will need one which reads up to 220°F (104°C). You may already have a candy-making thermometer, or you can buy specialist wax thermometers which cover the same temperature range.

FLAT BAKING TRAYS

Shallow baking trays can be used to hold molds while you pour in the wax, or in projects where you need to prepare a thin layer of wax.

MOLD RELEASE

Use mold release with rubber molds. Each application will last for about five pourings. You do not need mold release with rigid molds.

MOLD SEAL

This is a sticky, putty-like material used to seal the base of the candle mold to prevent molten wax from seeping out. It never goes hard and so you can use it time and time again. You can also use plasticine.

READY-MADE MOLDS

MOLD RELEASE

KNIFE

IMPROVIZED MOLDS

THERMOMETER

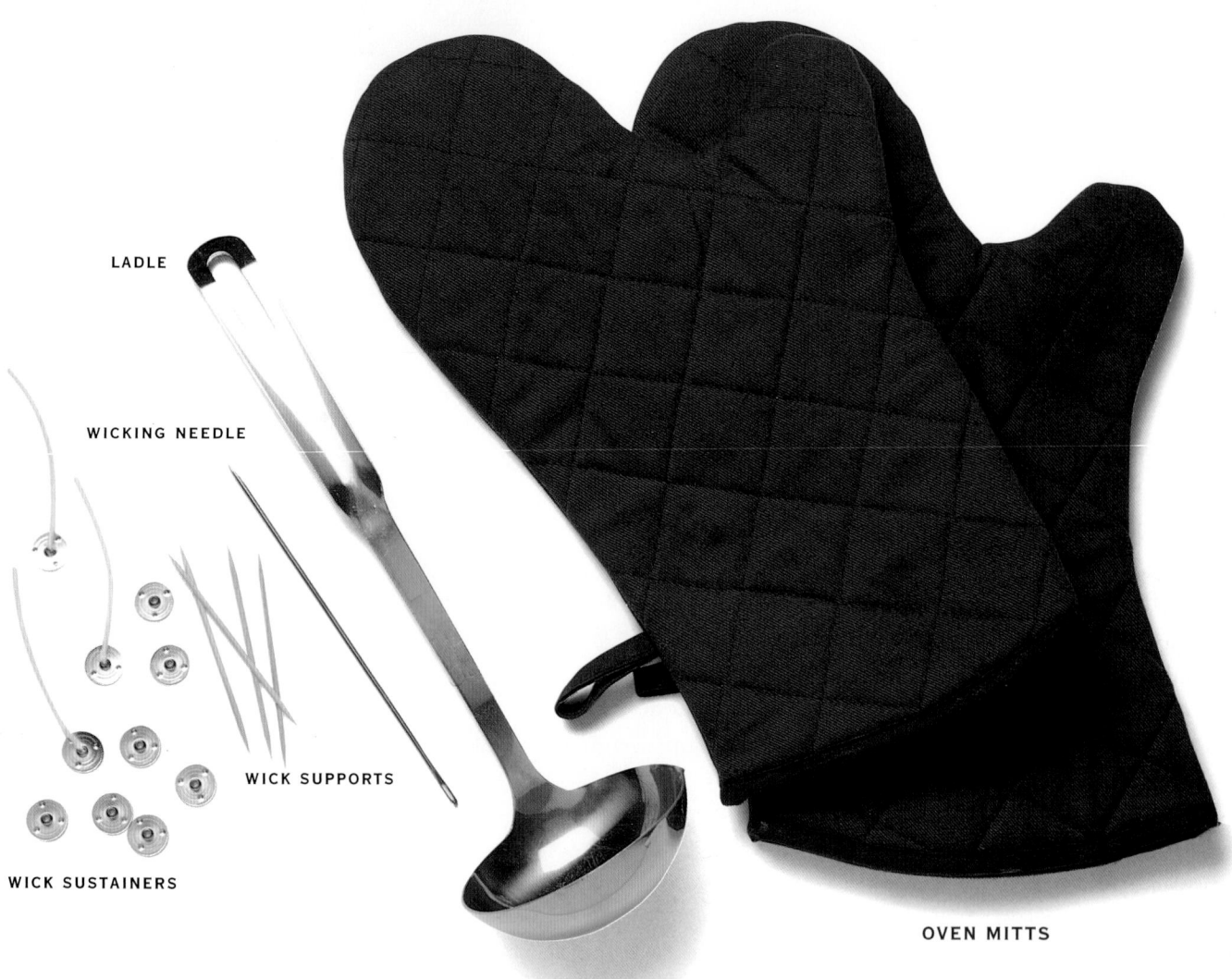

LADLE

WICKING NEEDLE

WICK SUPPORTS

WICK SUSTAINERS

OVEN MITTS

WICKING NEEDLE

This is a large, industrial needle available from candle-making suppliers and is available in lengths from 5 to 12in (13 to 30cm). It is used to thread the wick through the mold— threading a large latex mold without a wicking needle would be almost impossible. It can also be used as a support for the wick at the top of the candle. If you cannot find one, then use a large darning needle instead.

WICK SUSTAINERS

These are small metal disks which hold the wick at the bottom of container candles. They are either made separately or have the wick already attached.

WICK SUPPORTS

Cocktail sticks make good wick supports if the candle is not too wide. For wider candles, pencils or chopsticks are ideal.

OVEN MITTS

Keep a pair of old oven mitts especially for candle-making. Dipping cans and foil containers of molten wax can become quite hot.

KITCHEN SCALES

If your paraffin wax is in block form, you will need scales to work out the proportion of stearin (see page 13). If you are using powdered wax, you can measure by volume, i.e. nine spoonfuls of wax to one of stearin.

LADLE

A kitchen ladle is a useful way of transferring molten wax from one container to another.

DISHWASHING LIQUID

This will prevent wax from sticking to itself or to any other surface. It is used in many of the projects in this book.

WAXES AND DYES

*C*ANDLES ARE MADE USING WAX, WICKS, AND DYE. (WICKS ARE DEALT WITH IN THE NEXT SECTION.) THE SCIENCE OF CREATING A CANDLE WHICH WILL BURN CORRECTLY HAS BEEN PERFECTED OVER THE LAST CENTURY AND IT IS ESSENTIAL THAT YOU USE CUSTOM-MADE INGREDIENTS TO ACHIEVE GOOD RESULTS.

WAXES

Candles have not always been made from wax—in fact, the chief ingredient was once tallow (beef fat). Nowadays, waxes of different types are used.

PARAFFIN WAX is the most common wax used for candle-making, and is available in three forms: slab, pelleted, or powdered. You can buy paraffin wax with many different melting points. The one most usually chosen for candle-making has a melting point of 160°F (71°C).

STEARIN is a white powder which melts to a clear liquid when heated. It has four main functions: to help release the candle from the mold; to act as a hardener and enable the candle to burn longer; to dissolve dyes more thoroughly than they do in wax; and to give the candle a more opaque appearance.

It is not essential to use stearin. If you use a slightly thinner wick you can produce a candle that will burn perfectly well, if somewhat faster, without it. Indeed there are some instances when it is better not to use stearin, for example, when making candles in rubber molds. Generally, however, you will find it much easier to get a candle out of its mold if you include stearin. The proportion of stearin to wax is 1 to 9 (i.e. 10 per cent stearin).

MICROCRYSTALLINE SOFT is a wax with a very low melting point. It is ideal when you want to mold wax by hand and keep it soft.

MICROCRYSTALLINE HARD has a high melting point, which makes it much more resilient.

MICROCRYSTALLINE SOFT WAX

MICROCRYSTALLINE HARD WAX

POWDERED PARAFFIN WAX

SLAB PARAFFIN WAX

PELLETED PARAFFIN WAX

STEARIN

BEESWAX SHEETS

MODELING WAX

WAX GLUE

SOLID BEESWAX

DIP-AND-CARVE WAX is paraffin wax which is sold ready-mixed with microcrystalline soft wax. It is much more malleable than ordinary paraffin wax. It is used for dipping and carving candles because it does not split and crack as it is carved.

MODELING WAX is a soft, usually pre-colored wax. It can be used to make candles without being heated first.

SOLID BEESWAX is a natural product with a beautiful, honey smell. In its original form, it is a honey color, but you can also buy it bleached white or colored. It is perfectly suited for making hand-dipped candles. You can also use it very satisfactorily in latex molds. However, it is much more difficult to mold candles out of beeswax, even if the wax mixture contains only 10 per cent beeswax. This is because it is slightly sticky in character and it does not contract enough to remove the candle easily from its mold. Therefore, if you are using either pure beeswax or part beeswax to produce a rigid molded candle, lightly coat the inside of the mold first with dishwashing liquid to aid release.

BEESWAX SHEETS come in flat sheets, usually about 5 x 15in (13 x 38cm) with a tiny honeycomb pattern. Because beeswax is so soft and pliable, you can roll up these sheets and make beautiful candles without having to heat anything. It is now available in a range of lovely colors.

WAX GLUE is an extremely soft and sticky wax which is used to apply decorations to the outside of candles. It comes in small, solid blocks. To use it, simply smear a little either on the surface of the candle, or on the decoration. Then press the object gently and firmly onto the side of the candle. You can also use a little of it on the bottom of a candle to fix it firmly in its holder.

DYE DISKS

POWDERED DYE

DYES

It is important to use purpose-made candle dyes for the best results. Dyes come in different forms and it is important to know what you are buying. Most large candle manufacturers use pigments which do not fade. However, most of the colors available to the home candle-maker will fade if exposed to direct sunlight. It is amazing how fast a color can completely disappear from a handmade candle if it is placed close to a window.

Dyes come in two forms, either powdered or disk.

DYE DISKS are dyes which have already been dissolved in a little stearin. It is usually necessary to dissolve some of the disk in stearin before adding the wax, although some disks are now available that will dissolve directly in wax without being dissolved in stearin first. Make sure it is all thoroughly dissolved before you add the wax. Dye disks are better for beginners.

POWDERED DYES are extremely potent: a little goes a very long way. They need to be dissolved in stearin first. Most people graduate to powdered dye when they begin to make reasonable quantities of candles in batches of, say, 20 or so at a time.

STRENGTH OF DYES
Dyes have different strengths. Some powdered dyes need only about a matchhead of dye to color over 2lb (900g) of wax. It is best to start with a very little dye and gradually add more until you set the right color—it can take an enormous amount of wax to dilute an over-dyed batch.

Although the range of dyes available to the home candle-maker is wide you may want to mix your own colors. You can do this with all forms of dye.

WICKS

*W*ICKS ARE MADE FROM BRAIDED COTTON WHICH HAS BEEN TREATED WITH BORIC ACID. IT IS ESSENTIAL THAT YOU USE PROPER WICKS AND DO NOT TRY TO IMPROVIZE WITH STRING. NOT ONLY COULD THIS BE DANGEROUS, BECAUSE PIECES OF BURNING CARBON COULD FALL FROM THE CANDLE, BUT IT WILL ALSO MAKE YOUR CANDLES SMOKE.

TYPES OF WICK

Wicks come in different thicknesses, and should be chosen according to the final diameter of the candle you are making. When you buy wicks, they usually came labeled as ½in (1cm), 2in (5cm), etc. These measurements refer to the diameter of the candle for which they are suitable, assuming you are using candle-making wax with 10 per cent of stearin added.

CONTAINER WICKS

These are used in container candles. When a candle burns in a container, it can create quite a large pool of molten wax. When the candle is put out, the wick tends to flop over and disappear into the wax. For this reason, rigid wicks have been developed. They are either woven more tightly than usual, or stiffened with a plastic or cotton core. Until very recently lead was used, but this has now been banned in most countries.

PRIMED WICK is wick that has been dipped once in molten wax and then removed. This makes it stiff, which is useful for some candle-making methods. It also helps if the wick at the top of a candle has been primed, as this makes it easier to light.

PRIMING THE WICK

Heat the wax to 180°F (80°C). Take a length of wick and dip it once into the wax. Pull it straight and hang it up to dry—this only takes a couple of minutes. If you are using thick wick or sash cord make sure you leave the wick in long enough to completely absorb the wax.

BRAIDING WICKS

Buy your wicks wide enough for the candles you are making. The only exception to this is when making outdoor candles where you can create a large wick by plaiting smaller wicks together. Although this will cause the candle to smoke—which does not matter outdoors—it also means that the flame will be large enough so that the candle stays alight even in a strong wind.

TYPES OF WICKS

CHOOSING THE RIGHT WICK

Choosing the right wick size for your candles is essential if they are going to burn evenly and without smoke or dripping.

WICK FOR A MOLDED, UPRIGHT CANDLE

If you are going to make a candle which is 2in (5cm) in diameter, for instance, use a "2in/5cm" wick. The candle will then burn with a pool of wax which reaches just to the outside of the candle and remains the same size as the candle burns down (assuming that you are using paraffin candle wax with 10 per cent of stearin added).

WICK TOO LARGE

When a wick is too large, the candle will smoke. You can tell if a candle is smoking by the shape of its flame. If the tip is rounded, the candle is burning correctly. If flame flares out at the tip, the candle is smoking.

WICK TOO SMALL

This produces two different results, depending on the size of the wick. If it is only slightly too small, the candle will drip. If it is much too small—for example, a 1in (2.5cm) wick burning in a candle with a 3in (8cm) diameter—the flame will create a well in the center of the candle.

Sometimes a smaller wick is used on purpose, particularly when you want a candle to be refillable. This simply means allowing the candle to burn until a well has been created in the wax. You can then refill the well, using either some powdered wax and a piece of primed wick, or a night light candle.

PYRAMID- OR CONE-SHAPED CANDLES

Which wick should you choose for a cone- or pyramid-shaped candle— a thin wick, so the top burns correctly, or a thick one, so the bottom does? Choose a size that corresponds roughly to half the width of the base of the candle. The candle will then burn correctly at first, until about a third of the way down, when it will burn down leaving a shell.

CONTAINER CANDLES

Always use a much smaller wick in a container candle than you would in a free-standing one. Although it is not essential, you can buy rigid wick for use with container candles (see page 23).

You can buy wicks attached to round, metal wick sustainers, or you can buy wick sustainers and wicks separately. However, if you are unable to find a supplier, you can simply fix the wick to the base of the container by prodding it into a thin layer of half-set wax.

CENTERING THE WICK

It is very important that the wick is well centered, particularly in a glass container. If the flame gets too near the edge of the glass it will break. You can use the same method as you would for molded candles (see page 19) and secure the wick at the top of the container with a cocktail stick or pencil.

PREPARING THE MOLD

*T*HE VARIOUS TYPES OF MOLDS WILL NEED TO BE PREPARED IN DIFFERENT WAYS. BEFORE STARTING, CHECK ALSO THAT YOU HAVE THE RIGHT TYPE OF WICK FOR THE MOLD YOU ARE USING.

FLEXIBLE MOLDS

These include rubber and latex molds. Rubber is a natural substance, latex is its synthetic substitute. It is not necessary to use stearin in the wax mixture.

YOU WILL NEED

* * * * * *

Rubber or latex mold
*
Mold release
*
Hot, clear paraffin wax mixture
*
Wicking needle
*
Wick
*
Cocktail stick
*
Mold support
*
Baking tray

* * * * * *

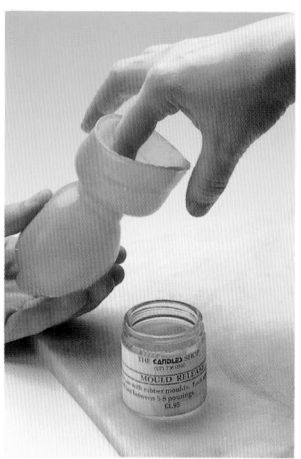

1 Using your finger, lightly coat the inside of the mold with mold release (each application of mold release will last for about five candle pourings).

2 Follow Steps 1–4 in the previous instructions with the exception of Step 2. You do not need to use mold seal with rubber or latex molds because the material grips the wick tightly and prevents the wax from seeping out.

* OPEN-ENDED * MOLDS

* When you use an
* improvized, open-ended
* mold, such as a cardboard
* tube or piece of drainpipe,
* keep it the right way up
* when filling it, rather than
* upside-down as with ready-
* made molds.
*

* YOU WILL NEED

* * * * * * *

Mold
*
Baking tray
*
Mold seal
*
Wick

* * * * * *

1 Fix the wick to the baking tray. You can do this by securing it with a lump of mold seal which you can cut out after-wards. Or pour a little wax onto the tray and push in the wick when the wax is almost hard.

2 Place the mold on the baking tray over the wick and press mold seal around its base.

3 Thread a cocktail stick through the wick and balance it across the mold. The tension should be tight enough to ensure that the wick is straight. Make sure the wick is placed centrally in the mold (see page 17).

RIGID MOLDS

These include readymade metal or plastic molds. For cardboard molds, see under *Open-ended molds* (page 18).

YOU WILL NEED

* * * * * * *

Wick
*
Hot paraffin wax mixture
*
Wicking needle
*
Candle mold
*
Mold seal or plasticine
*
Cocktail stick
*
Mold support
*
Baking tray or newspaper

* * * * * * *

TIP *Place the prepared mold in its container on some newspaper or a baking tray. Very occasionally, you may find that you have not pressed the mold seal down properly, so it is a good idea to be prepared for possible spillage.*

BEFORE YOU BEGIN

First decide the wick size you want to use (see *Choosing the right wick*, page 17). Dip the tip of the wick in the hot wax so it is primed. This prevents the mold seal or plasticine from getting into the wick.

1 Using a wicking needle, thread the wick through the mold. Most candles are made "upside-down" so you will be threading the wick through what will be the top of the candle. The primed section of the wick should be at the top end of the mold.

2 Take a lump of mold seal or plasticine and put it over the wick at the top of the mold. Make sure it is well pressed down around the edges.

3 Repeat step 3 as described for Open-ended molds.

4 Support the mold so that it is upright. You can use any suitable cylindrical object. Old tin cans or containers are perfect, as they will catch any wax that may seep from the mold.

SHALLOW MOLDS

If you are making a candle in a very shallow mold, you can put the wick in after pouring the wax. Simply prime the wick in some molten wax. Straighten it and let it harden. Then push it down into the wax before you top the candle up. If it flops down into the wax after topping up, just straighten it out again.

THE WAX MIXTURE

*A*LTHOUGH YOU CAN BUY WAX WHICH IS ALREADY MIXED WITH STEARIN, AND DYES WHICH YOU CAN PUT DIRECTLY INTO THE MIXTURE, THE FOLLOWING STEP-BY-STEP GUIDE ASSUMES THAT YOU HAVE BOUGHT YOUR WAX AND STEARIN SEPARATELY, AS THIS IS MORE USUAL.

PREPARATION

You will learn to judge how much wax you need but, as a rough guide, a candle about 6 x 2in (15 x 5cm) will take about 8oz (225g) of wax and about 1oz (25g) of stearin. When using a rubber mold, omit the stearin.

If you want to add perfume, see *Perfuming Candles* (page 26).

YOU WILL NEED

* * * * * * *

Stearin
*
Double boiler or substitute (see Step 1)
*
Sharp knife
*
Disk of candle dye, or powdered dye and spoon
*
Wax
*
Thermometer

* * * * * * *

1 Place the stearin in the top half of the double boiler (or in a container placed over a saucepan of hot water).

2 Heat until the stearin has completely melted (it changes from a white powder to a clear liquid).

3 Using a sharp knife, cut off a small piece of dye from a dye disk or, if you are using powdered dye, put a very small amount on a spoon.

4 Add the dye to the stearin and stir gently.

5 Add wax to the stearin and dye mixture.

6 Gently heat the mixture until it is molten. Check the temperature with a thermometer. It is ready to pour when it has reached 180°F (80°C).

* POURING

* Make sure that the jug is
* not too cold as this will
* lower the temperature of
* the wax. Candles poured
* at below 180°F (80°C)
* will be difficult to get
* out of the mold and may
* be spoilt by white marks.
*
1 Ladle about two-thirds of the prepared wax mixture into a pouring jug.

2 Gently pour in the wax up to about ¼in (5mm) from the top of the mold.

COOLING A CANDLE

Candles which set in cool conditions harden more quickly and have a shinier finish than those poured in room temperatures above 65°F (18°C). For this reason some people place their candles into a water bath for cooling.

Topping up is done while the candle is in the water. However, water-cooling can create more problems than it solves. If water gets into the mold as the wax is setting, the candle will not burn properly. Also, if the waterline is below the height of the wax in the mold, white marks may form at the water level.

Unless you are making candles in the height of the summer, or in a tropical climate without air conditioning, it is certainly not necessary to cool the candle in this way. If the candle is not as shiny as you would like, you can always give it a final dip (see *Finishing Touches*, page 25).

TOPPING UP A MOLD

As a candle cools, it contracts. The degree of contraction will depend on the size of the candle, but will usually be around 25 per cent of the volume. The top of the candle will dip and form a well and there will be air bubbles hidden below the skin of the wax.

This is why the topping-up procedure is essential for molded candles. If you do not follow it, your candles may look all right on the outside, but inside they will be full of holes and may capsize while burning. In addition, the shape of candles poured into rubber molds may be distorted by the wax contraction. (See also *Topping up* under *Container Candles*, page 23.)

TIP *If you are making a very shallow candle, up to 1in (2.5cm) or so deep, it will not be necessary to pierce the surface. If you are making a candle in a bowl or similarly shaped mold, you could add more variety by pouring your final top-up in a different color (see Scented Floating Candles, page 54).*

1 When the candle has developed a skin about ⅛in (3mm) deep, take a pencil or chopstick and prod the surface around the wick. This will ensure that the next pouring of wax will reach the holes that have formed below.

2 Heat up about half your remaining wax to 180°F (80°C) and ladle it into a pouring jug.

3 With great care, pour the wax into the hollow at the top of the candle. Make sure that it does not rise above the original level of the candle, or the molten wax may get between the setting candle and the mold. If this happens, the candle may not come easily out of the mold.

4 Top up several times, if necessary. When the wax has completely contracted, it will be flat on the surface of the candle.

REMOVING THE MOLD

*T*HE WAY YOU REMOVE THE CANDLE FROM ITS MOLD WILL DEPEND ON THE TYPE OF MOLD YOU ARE USING. ALWAYS REMOVE THE CANDLE CAREFULLY OR YOU MAY DAMAGE IT. IF YOU HAVE DIFFICULTY REMOVING IT, REFER TO *PROBLEMS AND REMEDIES* ON PAGE 29.

RIGID AND OPEN-ENDED MOLDS

This section covers all types of rigid molds.

FLEXIBLE MOLDS

This covers both rubber and latex molds.

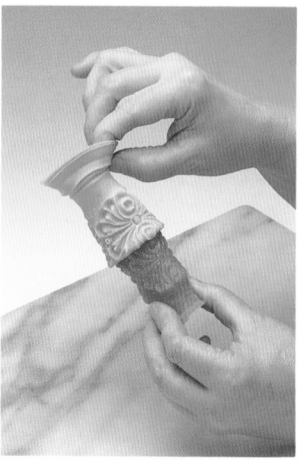

1 Remove the mold seal from the wick. For open-ended molds, remove the mold seal from around the base of the mold.

2 Holding the wick left at the base of the candle (or top in the case of open-ended molds), gently pull the candle out of its mold.

1 Smear a little dishwashing liquid on the surface of the mold. This is to prevent the rubber sticking when it is pulling back on itself. Do not use too much of the liquid, or the mold will be too slippery to grip properly.

2 Pull the rubber back over itself to expose the candle. In candles with very clearly defined shapes, this may be quite difficult at first, but it does become easier with practice.

CONTAINER CANDLES

$\mathcal{S}$EVERAL OF THE PROJECTS DESCRIBED IN THE BOOK ARE MADE BY THE SIMPLEST METHOD OF ALL—POURING WAX INTO CONTAINERS. CONTAINER CANDLES ARE VERY EASY TO MAKE, NOT LEAST BECAUSE YOU DO NOT HAVE TO WORRY ABOUT THE FINAL FINISH OF THE CANDLE OR WHETHER IT WILL COME OUT OF ITS MOLD. HOWEVER, THEY DO REQUIRE PARTICULAR CANDLE-MAKING SKILLS BECAUSE THE CONTAINER USED WILL AFFECT THE WAY IN WHICH THEY BURN.

PREPARING THE WAX MIXTURE

See *The Wax Mixture* (page 20).

CHOOSING THE WICK

As always, it is important that you select the right wick size. If the wick is too small the candle will just burn down the middle, leaving a rim of unburnt wax. If it is too large, it will smoke. Getting the wick size right is largely a matter of experiment (see *Choosing the Right Wick*, page 17).

INSERTING THE WICK

You can place the wick in its sustainer directly on the base of your container, even if it is glass. When the candle burns to the bottom, the heat of the flame will not crack it. However, the base of the candle may become rather hot and you may prefer to pour in a small amount of wax and let it harden before you insert your wick, or wick and wick sustainer. As with the *Scented Floating Candles* on page 54, you can also insert primed wick just before you top the candle up if the container you are using is very shallow.

TOPPING UP You need to top up a container candle just as you do a molded candle. However, it is much simpler to do, as it does not matter if you pour the top-up wax over the original level of the candle. (See *Topping up a mold*, page 21.)

CENTERING THE WICK

See *Wicks* (page 17),

HAND-DIPPED CANDLES

$\mathcal{D}$IPPING IS A TRADITIONAL METHOD OF MAKING DINNER CANDLES, AND HAS THE ADDED ADVANTAGE THAT YOU CAN MAKE CANDLES IN THE EXACT DIAMETER YOU NEED TO FIT YOUR OWN CANDLE HOLDERS.

YOU WILL NEED

* * * * * * *

Dipping can at least 12in (30cm) deep

*

Container that holds at least 12in (30cm) cold water

*

10lbs (4.5kg) of clear paraffin wax and stearin mix

*

12in (30cm) dipping can with any colored paraffin wax you like, with 1% microcrystalline wax added (optional)

*

22in (56cm) length of 1in (2.5cm) wick

* * * * * * *

TIP *To give the candle a hard outer layer and make it more drip resistant, add microcrystalline hard wax to your final color. You can, of course, make hand-dipped candles the same color all the way through. Try dipping in several colors for a rainbow effect.*

1 Cut off a 22in (56cm) of 1in (2.5cm) wick. Drape it over the four fingers of one hand so that the two sides are of equal length. Keeping the two strands of wick apart during dipping is essential, or the candles will stick together.

2 Prepare a dipping can of clear wax heated up to 170°F (76°C). Lower the wicks into the can until the wax almost reaches your fingers. Lift the wicks up gently. The dipping movement should be done as smoothly as possible or the candles will develop lumps. Always dip right to the top of the candle – the familiar tapered shape forms naturally as the wax falls down the candle each time it is lifted out.

3 Dip the wicks into the container of cold water.

4 Repeat Steps 2 and 3 until the candle reaches a diameter of just under an inch (2.5cm).

5 Fill a dipping can with hot water. Pour in a 1in (2.5cm) layer of colored wax. The temperature should be at least 180°F (80°C). Dip the pair of candles in the colored wax. You will probably need at least two further dips to achieve a strong color.

FINISHING TOUCHES

*T*HIS CAN BE ONE OF THE MOST ENJOYABLE AND REWARDING PARTS OF CANDLE MAKING. IT'S THE POINT AT WHICH YOU CAN STAMP YOUR OWN PERSONALITY ON YOUR CREATIONS!

MOLDED CANDLES

The look of your finished candle will depend very much on the type of mold used and the temperature at which the wax was poured and cooled. If you have used a ready-made rigid plastic or metal mold, ensured that the pouring temperature was correct, and have allowed it to set in a cool environment, you will be left with a shiny, perfectly finished candle. However, don't despair if the candle looks rather dull when it comes out of the mold—either of the techniques below will improve its appearance.

BUFFING IT UP

Using a soft cloth, gently rub the candle until its surface becomes shiny. This is particularly useful for candles which have been poured into latex molds.

OVER-DIPPING

Over-dipping a candle will camouflage a multitude of sins. You can achieve this in two ways:

✳ *Add a layer of wax* by dipping the candle into a can containing wax heated to 180°F (80°C). If you want to keep the original color, dip it in clear wax, or change the color by dipping it into highly dyed wax of a different hue. If you then dip it into water immediately, you will achieve an even greater shine. Make sure that the water is not too cold or the new surface layer of wax will crack.

✳ *Remove a layer of wax* by dipping into a can containing wax heated to 190°F (87°C). This is particularly useful if you are making a candle which has wax chips or shapes which you want to expose (see *Decorative Pyramid*, page 116).

PAINTS AND LACQUERS

There are particular points to bear in mind when painting candles. You may find it takes a little time to develop your technique, even if you are an experienced painter.

PAINTS

It is not possible to use paint on its own directly onto a candle as the paint does not adhere to the wax. Dilute either poster or acrylic paint with dishwashing liquid. Take into account that the paint will take longer than usual to dry, sometimes as long as 24 hours.

CANDLE LACQUER

This is a varnish which gives a candle a very high shine. Do not use any other varnish as it may not be suitable, and may be dangerous. Candle lacquer is highly inflammable and should be used away from heat.

METALLIC CANDLE LACQUER

This lacquer has been specially designed to adhere to wax and to burn safely. It cannot be painted directly onto candles, but you can dip candles into it.

GOLD WAX

Rubbed onto the surface of a candle, gold wax looks particularly good when used to highlight an embossed candle.

PERFUMING CANDLES

Most people enjoy the scent of a beautifully perfumed candle. Candles are as efficient as an atomizer for dispersing perfume into the air. In an atomizer, the perfume vaporizes with the hot oil. The same thing happens with candles. As the candle burns, a pool of wax forms on its surface. This hot wax produces a vapor and the perfume is released. The greater the area of molten wax, the more efficient will be the dispersion of the perfume. Container candles are particularly effective at spreading perfume—they create a larger pool of molten wax than a block candle. A perfumed dining candle will be pleasant, but really no more effective than using potpourri since the molten area is so small.

CANDLE-MAKING PERFUMES

Use perfumes which have been mixed especially for use in candles. If you use other perfumed oils they may affect the burning quality of the candle.

Candle scents vary greatly in their potency. How much perfume you put in a candle depends very much on your own personal taste and the quality of the perfume you buy. Perfumes vary greatly in quality. Some are made from natural oils while others are made from much cheaper chemical equivalents. You usually get what you pay for. Most candle-makers' suppliers will give you a rough guide to their use on the bottle, but do experiment.

PERFUMES AND THEIR PROPERTIES

Perfumes produce different effects. For instance:

Bergamot for uplifting your spirits

Ylang ylang for romance

Rosemary to stimulate and energize

Lavender for relaxation

Frankincense for rejuvenation.

All candles will clear the air of cigarette smoke, but try adding vanilla perfume—it really does banish the smell of stale tobacco. Citronella-scented candles will help to discourage insects. Citronella has long been used as a really effective insect- and cat-repellent by those trying to enjoy a warm evening outdoors.

MACHINE-MADE CANDLES

*I*T IS USEFUL TO BE ABLE TO TELL HOW CANDLES PRODUCED COMMERCIALLY HAVE BEEN MADE. IF YOU ARE DECORATING OR CARVING READY-MADE CANDLES, THE ORIGINAL METHOD USED TO MAKE THE CANDLE MAY AFFECT YOUR RESULTS.

DRAWN CANDLES

Most church candles are now made by this method. It involves passing a very long piece of wick through a trough of wax over and over again until the desired thickness has been built up. The candles are then cut to size and milled. The effect is very much the same as for hand-dipped candles except that drawn candles are straight-sided and not tapered.

MOLDED CANDLES

A roll of wick is kept under each mold so that when the set candle is pulled from the mold, the wick can be drawn up into the now-empty mold ready for the next candle pouring. This is known as automatic wicking. Many factories still make molded candles. The molds are usually arranged together in banks.

COMPRESSED POWDERED WAX

Many block and floating candles are made by machines which compress powdered wax into shape. If you are carving into an over-dipped candle of this type, you may find that it crumbles a little.

EXTRUDED WAX

This method involves pulling half-molten wax through a machine—rather like squeezing toothpaste from a tube. Traditional dinner candles are made in this way.

HAND-DIPPED CANDLES

There are many factories, particularly in Scandinavia, where candles are hand-dipped. Hundreds of pairs at a time are dipped on frames. Candles made in this way are not suitable for carving as they tend to flake off in layers when cut.

WORKING SAFELY

Candle-making is a safe and enjoyable hobby. However, molten wax needs to be treated with the same respect as hot fat or oil. If candle-making wax is heated above 212°F (100°C), it will ignite.

FIRE PRECAUTIONS

By following these simple precautions, you will ensure that you never have a fire.

✳ Never leave heating wax unattended.

✳ Use a double boiler, or a metal or foil bowl over a saucepan of water, to heat the wax. If you are placing a container of wax directly into a saucepan of water, put a trivet in the bottom of the pan to lift the container and ensure that it is not directly above the heat source. By heating the wax in this way, it cannot overheat, even if you are momentarily distracted.

✳ Make sure the water in the saucepan never boils away.

✳ If you have to heat the wax to more than 180°F (80°C), use a thermometer. Put the thermometer in the wax as soon as it is molten and leave it in the wax until it has reached the required temperature.

✳ If you do have a wax fire, do not use water to extinguish it. Treat it just as you would a fat fire and put it out with a damp cloth, saucepan lid, or fire blanket. Or use a powder fire extinguisher.

✳ When designing candle holders, use non-inflammable materials. Space the candles well away from decorations, especially when using natural objects, such as pine cones or other dried materials.

✳ Never leave a burning candle unattended.

CLEANING YOUR EQUIPMENT

In most cases, wax contracts sufficiently to come easily out of any container it has been poured into. If you are left with a small residue, wash the container in very hot water and wipe it with a cloth.

SPILT WAX

You are bound to spill a little wax when you make candles. Place a small container below while pouring to minimize drips. Or cover your work surface with old newspapers. If you do spill some molten wax on the work surface, allow it to harden and then scrape it off using a soft spatula or your fingernail.

Clear wax is easily removed, but some dyed wax may leave a stain. If you cannot remove the stain, try rubbing it with a little white spirit.

Always wear an apron when candle-making. If you spill wax on your clothing, it can be removed quite easily. Place some newspaper or absorbent tissue paper on the wax spot and run a warm iron over it. The wax will be absorbed by the paper. Move the newspaper and repeat until no more wax appears.

LEFTOVER WAX

Never pour any molten wax down the sink. It will harden and be very difficult to remove.

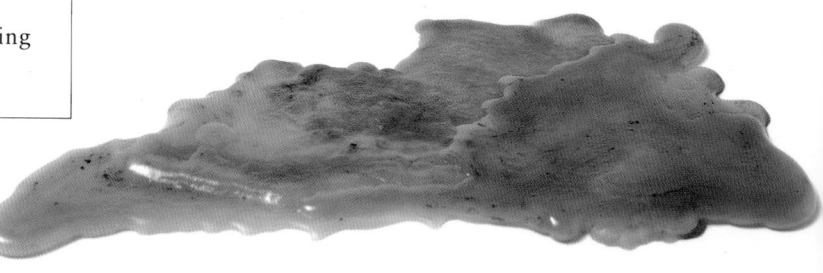

PROBLEMS AND REMEDIES

EVERYONE MAKES MISTAKES WHEN MAKING CANDLES—IT'S PART OF THE LEARNING PROCESS! THE TABLE BELOW GIVES WAYS OF DEALING WITH THE MISTAKES WHEN THEY HAPPEN. EVEN IF A CANDLE HAS GONE IRRETRIEVABLY WRONG, YOU WON'T HAVE WASTED MATERIALS, SINCE MOST OF THE INGREDIENTS ARE RE-USABLE.

THE CANDLE WILL NOT COME OUT OF THE MOLD

This is probably because:
* you poured the wax in too cold—wax poured below 180°F (80°C) may not contract sufficiently to enable the candle to fall out; or
* some wax spilt over the original level of the candle when you topped it up, and formed a wedge between the candle and the side of the mold.

Put it in the refrigerator for a couple of hours. If it still will not come out, re-heat the mold in hot water until the candle slides out. Unfortunately this will ruin the surface of the candle, and you will have to clean the mold. You can use the wax again.

THE CANDLE SPITS WHILE BURNING

Water got into the candle while it was being water-cooled.

THE CANDLE IS COVERED WITH WHITE, HORIZONTAL LINES

When wax is poured in at below 160°F (71°C), it can form white lines as it cools against the side of the mold.

You can either leave them—some people enjoy them as a special effect— or remove them by dipping the candle in wax heated to 190°F (87°C).

THE CANDLE FORMS WHITE BUBBLES AFTER IT HAS BEEN OVER-DIPPED

This happens when the over-dip is done with wax that is too cold. Air gets trapped between the candle surface and the over-dip.

Hold the candle in a pan of wax heated to 180°F (80°C) and leave it there until the outer layer has melted. Then take it out, wait for it to harden again, and over-dip once more, making sure that this time the wax is hot enough, i.e. over 180°F (80°C).

THE CANDLE HAS TINY, WHITE, SMUDGY BLOTCHES

This has nothing to do with your technique. It is because some waxes are refined better than others and you can be unlucky and get a faulty batch.

If you do not like the effect (it can look quite pretty), camouflage the candle by over-dipping it in a different-colored wax. To prevent this happening with the rest of this particular batch of wax, try adding a tiny proportion of micro-crystalline hard wax.

THE CANDLE IS COVERED WITH TINY PINPRICK HOLES

When wax over 190°F (87°C) is poured into the mold, it forms tiny bubbles as it cools. This cannot happen when you are using a thermometer, or heating wax over water, and the fact that it has happened shows that you are not working safely!

THERE ARE VERTICAL CRACKS IN THE TOP OF THE CANDLE

The wax you used to top up the candle was too hot—well over 180°F (80°C).

METALLIC CANDLES

PIN CANDLE

YOU WILL NEED

* * * * * * *

Large, cream-colored candle

*

Ruler

*

Gold upholstery pins

* * * * * * *

*T*HIS DRAMATIC FINISH IS SO SIMPLE TO ACHIEVE — TRY EXPERIMENTING WITH DIFFERENT SHAPED PIN HEADS AND DESIGNS.

3 Placing your ruler against the two completed circles, insert a pin exactly halfway down the candle. Then repeat the stages illustrated in Steps 1 and 2 once more.

2 Remove the ruler, and insert more pins in a circular shape all around the top pin. Repeat with the bottom pin.

1 Place a ruler against the candle, making sure it is parallel with the edge. Press in one pin 1½in (4cm) from the top, and another 1½ (4cm) from the bottom.

*T*HE DECORATION CAN BE SIMPLE OR COMPLEX. VARY THE RESULT BY ADDING GOLD-PAINTED OR PAPER BACKING TO THE PINS.

STARRY CANDLE

YOU WILL NEED

* * * * * * *

Gold braid

*

Glass tumbler

*

Wax glue

*

Mixing jug

*

About 4oz (125g) powdered paraffin wax

*

Small gold paper or polythene stars

*

Primed wick and wick sustainer

*

Scissors

* * * * * *

$\mathcal{A}$NOTHER PROJECT THAT'S FUN TO DO WITH CHILDREN. EXPERIMENT WITH DIFFERENT CONTAINERS — OR USE MORE THAN ONE WICK IF YOU HAVE A WIDE GLASS BOWL HANDY.

2 In the jug, mix some powdered wax with the tiny gold stars so they are evenly distributed throughout the wax.

TIP *The beauty of powdered wax is that you can use it to fill any container without having to heat the wax first.*

1 Cut out two pieces of gold braid about ⅛in (3mm) longer than the circumference of the tumbler. Wrap them around the glass and fix in place with a tiny blob of wax glue.

3 Place the primed wick and a wick sustainer into the glass and spoon in the powdered wax around it.

4 Make sure the wick is centered and trim about ½in (1cm) above the level of the wax.

$\mathcal{V}$ARY THE SHAPE AND DECORATION OF THE CONTAINER. USE CLEAR OR COLORED GLASS TO PRODUCE A RANGE OF EFFECTS.

GOLDEN BALL CANDLE

YOU WILL NEED

* * * * * * *

*F*RAMING THE
GOLDEN BALL CANDLE
IN A DELICATE FAN
OF WAX MAKES A
QUITE ORDINARY
CANDLE APPEAR
WONDERFULLY
EXOTIC.

4oz (125g) paraffin wax

*

*2 oz (50g) microcrystalline
soft wax*

*

Ivory dye (see page 15)

*

*Baking tray, smeared with
dishwashing liquid*

*

*Round object to use as
template, such as a glass or
jar lid, approximately twice
diameter of candle*

*

Craft knife

*

*Ball candle, coated with
gold lacquer*

* * * * * * *

1 Prepare a mixture of
paraffin wax and 20%
microcrystalline soft
wax, colored with a
small amount of ivory
dye. Heat to 180°F
(80°C). Pour one ladleful
into the tray, and allow it
to spread out.

3 Remove the excess
wax from around the cut-
out circle.

2 When the wax has
turned opaque and is soft
but no longer liquid,
place the round object on
the wax. Cut around it.

4 Carefully lift the wax
circle, and place the ball
candle in the center.

TIP *It is much easier to
create an attractive shape
if the sheet of wax you
are using is really thin.
Practice thinning the wax
out with a ladle, as if
you were making a
pancake, before you let it
cool and cut out the shape.*

5 Holding the wax circle and candle in one hand, use the fingers and thumb of your other hand to crinkle and fan out the edge of the circle.

A GROUP OF THESE CANDLES PRODUCES A RICH AND MAGICAL EFFECT.

MIRROR CANDLES

YOU WILL NEED

* * * * * *

MIRRORS AND CANDLES WORK WONDERFULLY TOGETHER. THE MIRROR MAGNIFIES THE LIGHT FROM THE CANDLES AND, HERE, REFLECTS THE GOLD.

2 gold ball candles approximately 2in (5cm) in diameter

*

1 gold ball approximately 3in (7.5cm) in diameter

*

Oblong mirror tile, 5 by 10in (12 by 25cm)

*

Mold seal

*

Holly leaves

*

Nuts

*

Gold spray paint

*

English ivy leaves

* * * * * *

TIP *Experiment with different shapes of candle and tile, and different decorations.*

THE RANGE OF DECORATIVE MIRRORS NOW AVAILABLE MAKES THE POSSIBILITIES OF THIS PROJECT ALMOST INFINITE. EXPERIMENT WITH DIFFERENT COLORED AND SHAPED CANDLES.

3 Spray the holly leaves and nuts with gold paint. Press them in place along the edges of the tile. Add a few green ivy leaves, for colour contrast.

2 Press mold seal neatly along all four edges of the tile.

1 Arrange the three ball candles on the tile.

ICE CANDLE

YOU WILL NEED

* * * * * * *

THIS TECHNIQUE QUICKLY TRANSFORMS AN EVERYDAY CANDLE INTO SOMETHING EXOTIC. THE ICE MAKES IT SET VERY QUICKLY AND IT WILL BE READY TO TAKE OUT OF THE MOLD IN UNDER AN HOUR.

White block candle 2 by 7in (5 by 18cm)

*

Candle mold at least 3 by 8in (7.5 by 20cm)

*

Mold seal

*

About 20 ice cubes

*

8oz (225g) wax mixed with 2oz (50g) stearin

*

Rub-on gold paint

* * * * * * *

2 Pull the wick tight and seal around it well with mold seal.

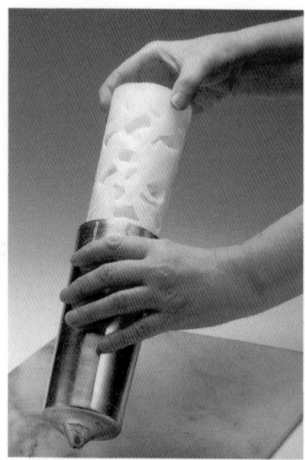

4 Pull the candle from the mold. It should drop out very easily.

TIP *A larger proportion of stearin makes the candle very opaque and brilliant white. It also makes it very easy to remove the candle from its mold.*

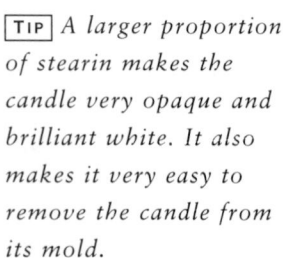

1 Place the candle into the candle mold and thread its wick through the hole at the top.

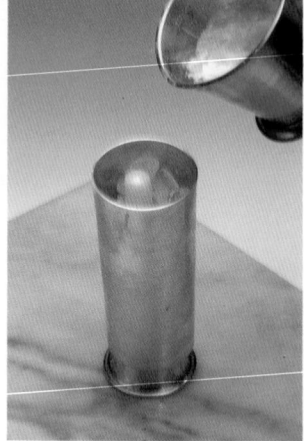

3 Turn the mold upside down. Making sure the candle stays in the center, drop ice cubes into the mold around the candle. Pour in the wax and stearin mixture so that it just covers the base of the candle.

5 Shake the candle to remove any water left by the molten ice cubes. Use the gold paint to outline the holes.

$\mathscr{P}$AINTED OR PLAIN, THE
EVER-VARYING PATTERNS
OF THIS CANDLE WILL
ALWAYS PROVIDE AN EASY
TOPIC OF CONVERSATION!

GOLDEN FRUIT BASKET

YOU WILL NEED

* * * * * * *

Transform garden mesh into a sophisticated candle holder. Either buy candles ready made, or make your own.

Sheet of galvanized steel wire mesh

Wire cutters

Gold ribbon

4 gold fruit candles

* * * * * * *

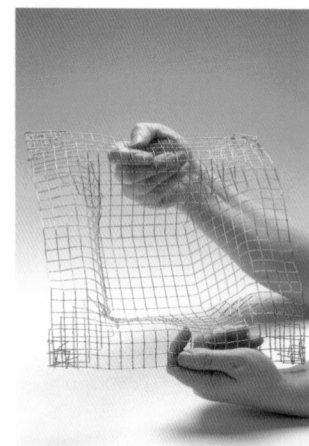

3 Push the sides of the square inward to form the edges of the bowl. Finally bend the edge of the rim downward. Make sure that the rim is an even width all the way round.

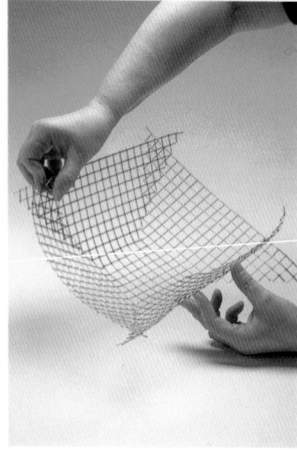

2 Carefully bend the mesh up on all four sides of the square. You will be left with a 1½in (4cm) overlap at each corner. Bend the overlaps back so that the corners are square.

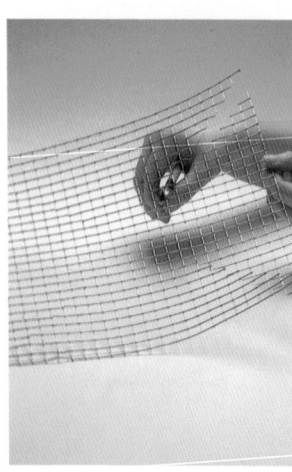

1 Use the wire cutters to cut out a 14in (35cm) square of the wire mesh. Then make four diagonal cuts from each corner into the center, leaving a 3in (7.5cm) square of uncut netting in the middle.

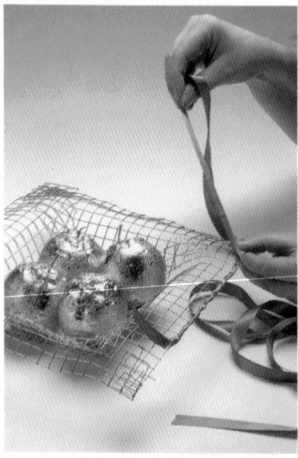

4 Wind gold ribbon around the edge of the bowl. Place four gold fruit candles (see page 46) in the middle.

TIP Wire mesh makes an ideal base to which you can add further decorations if you wish.

Gold is the theme here, but a different effect could be made with colored fruit and wax flowers.

AROMATIC CANDLES

LEMON CANDLE

*F*RUIT CANDLES CAN BE MADE TO LOOK SO REALISTIC THAT THEY LOOK GOOD ENOUGH TO EAT. DIFFERENT TYPES OF RUBBER FRUIT MOLDS ARE AVAILABLE, APART FROM THE LEMON USED HERE.

YOU WILL NEED

* * * * * *

Rubber lemon-shaped mold

*

Mold release

*

Length of 1½in (4cm) wick

*

Wicking needle

*

Container to hold mold, such as a tumbler

*

About 4oz (125g) yellow paraffin wax with no added stearin

*

Dishwashing liquid

*

Soft cloth for buffing wax

* * * * * *

2 Using a wicking needle, thread the wick through the center of the mold (see page 18).

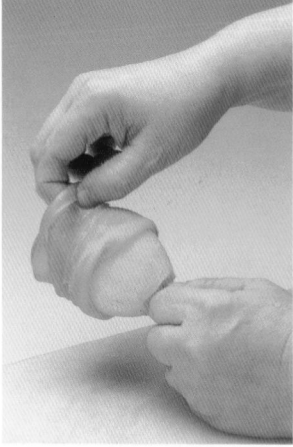

4 To remove the candle, smear the outside of the mold with dishwashing liquid so that it slides back easily on itself. Firmly holding the wick at the base of the candle, pull the mold back.

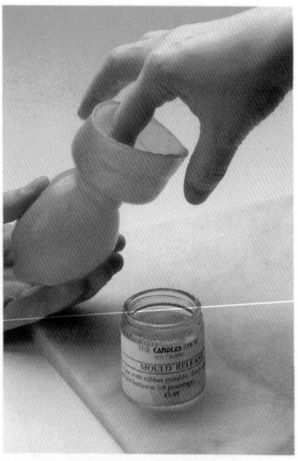

1 Using your finger, smear some mold release onto the inside of the rubber mold.

3 Rest the mold in the container. Prepare the yellow wax and pour into the mold. When the wax has almost cooled, prick the surface and top up with additional wax.

5 Buff the candle with the cloth to give a realistic shine. Other fruit candles may be made in the same way, using appropriate molds and colored wax.

TIP *Your rubber molds will last longer if you use very little or no stearin in your wax mixture.*

*F*RUIT-SHAPED CANDLES
CAN BE SO REALISTIC, THAT
IT'S EASY TO MISTAKE
THEM FOR THE REAL THING!

SPICY CANDLE

YOU WILL NEED

* * * * * *

THREE SPICES COMBINE TO MAKE A VISUALLY APPEALING AND EXOTIC-SMELLING CANDLE HOLDER. ANOTHER ATTRACTIVE VARIATION IS TO USE DRIED RED CHILLIES.

Glass tumbler

Bay leaves

Wax glue

Cinnamon sticks

Vanilla pods

Rubber band

Brown string

Candle to fit inside the glass tumbler

* * * * * *

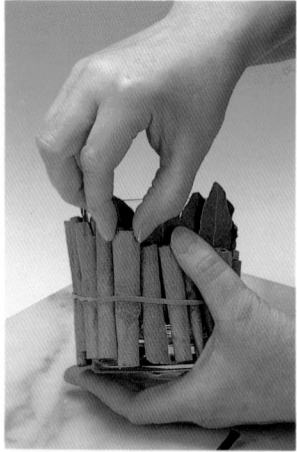

2 Trim the cinnamon sticks and vanilla pods if they are longer than the side of the tumbler.

3 Place the cinnamon sticks around the bay leaves, using the rubber band to hold them vertically in position.

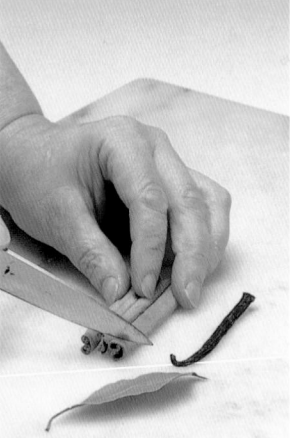

1 Using wax glue, press the bay leaves in a circle around the outside of the glass below the rim.

4 Slide a vanilla pod between about every fourth cinnamon stick.

5 Tie a double length of brown string just above the rubber band. Make sure that all the vanilla pods are vertical. Then tie a second piece of string below the rubber band.

6 Place the candle in the tumbler.

TIP *If you cannot find a candle that fits the tumbler of your choice, a night-light candle will work perfectly well. Its heat will still accentuate the scents of the herbs and spices.*

HERE THE SPICE DECORATIONS HAVE BEEN TIED ON FOR TEMPORARY USE, BUT YOU MAY WANT TO CREATE A MORE PERMANENT CANDLE HOLDER BY GLUING CARDOMOMS OR STAR ANISE TO YOUR CHOSEN CONTAINER.

AROMATHERAPY CANDLES

YOU WILL NEED

* * * * * * *

A SET OF AROMATHERAPY CANDLES — ONE FOR LOVE, ONE TO UPLIFT YOUR SPIRITS AND ONE TO RELIEVE STRESS! THEY MAKE WONDERFUL AND INEXPENSIVE PRESENTS FOR YOUR FRIENDS TOO.

About 1½lb (700g) dyed paraffin wax and stearin mix

*

12 drops of bergamot candle scent

*

12 drops of patchouli candle scent

*

12 drops ylang ylang candle scent

*

3 container wicks with sustainers

*

3 glass containers

* * * * * * *

2 Fill each glass with a different perfumed wax to within about 1in (2.5cm) of the top. Wait until the wax has almost set and then pull the wick to the center of the glass.

TIP *Making container candles in this way is quicker than fixing the wick centrally with a cocktail stick. However, do not try it unless you can keep an eye on the candles as they are setting. If the wax becomes too hard you will not be able to pull the wick to the center.*

1 Prepare three separate batches of wax and stearin mix. Add a different perfume to each batch. Put a wick sustainer in each glass container and pour in a little wax over it. Wait until it has almost set.

3 Top up each candle with the same wax you used for the main pouring.

*V*ary the color and scent of these slow-burning candles to suit your mood.

SMOKE-REPELLING CANDLE

YOU WILL NEED

* * * * * * *

ALL CANDLES WILL HELP CLEAR CIGARETTE SMOKE, BUT A CANDLE MADE ESPECIALLY FOR THIS PURPOSE MAKES A LOVELY GIFT. CHOOSE THE PERFUME TO SUIT THE TASTE OF THE PERSON YOU ARE GIVING IT TO.

About 4oz (125g) very pale green paraffin wax and stearin mix

*

2 drops of smoke-repelling perfume (see page 26)

*

Small silver container

*

2in (5cm) length of primed 1in (2.5cm) wick (see page 16)

* * * * * * *

1 Heat the wax and add the perfume. Pour it into the silver container to within ¼in (5mm) of the top.

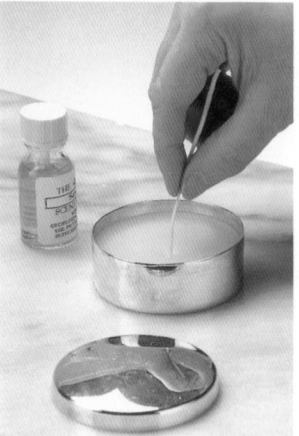

2 Wait until the wax has almost set but is still soft in the middle. Then press the primed wick into the center. When you feel it touch the base of the container, pull the wick out by about ⅛in (3mm). This will prevent the metal getting too hot when the candle burns to the bottom.

3 Top up the candle and cut the wick to within ½in (1cm) of the wax.

THIS CAN BE A TACTFULLY UNDERSTATED OBJECT, AS SHOWN HERE. ADD A VERY FEW DROPS OF CANDLE PERFUME — VANILLA IS IDEAL.

SCENTED FLOATING CANDLES

YOU WILL NEED

* * * * * * *

Colored paraffin wax and stearin mix

*

Perfume

*

Candle molds, such as a muffin tin or small, foil cake cases

*

Primed wick (see page 16)

*

Scissors

* * * * * * *

*T*HE SUBTLE REFLECTIONS CREATED WHEN SEVERAL FLOATING CANDLES ARE BURNING IN WATER ACCENTUATE THE ROMANCE OF CANDLELIGHT.

TIP *Floating candles are fun and easy to make as they set very quickly. They are also a great way of using up odd scraps of colored wax left over from other candles.*

1 Prepare the wax, mixing in a tiny drop of perfume. Pour the wax mixture into the molds.

2 When the wax has almost set but is still slightly soft, gently push a length of wick into the middle of each candle.

3 Leave the candles until they are almost completely set. Top up each candle using a different-colored wax. Wait for the candles to set fully, then remove from the molds. Trim the wicks to within ½in (1cm) of the wax.

*F*LOAT AWAY IN DREAMS WITH THESE TINY JEWEL-LIKE LIGHTS.

THREE-WICKED CANDLE

YOU WILL NEED

* * * * * * *

A WONDERFUL
REFILLABLE SETTING
FOR VANILLA-
SCENTED WAX. THE
THREE WICKS WILL
CREATE A POOL OF
MOLTEN WAX WHICH
WILL FILL THE ROOM
WITH PERFUME.

30 drops of vanilla scent

*

*About 1lb (450g) cream-
colored paraffin wax and
stearin mix*

*

Silver or other bowl

*

*3 lengths of ½in (1cm)
primed wick (see page 16)
and wick sustainers*

*

2 chopsticks

*

*Tube of silver stained-glass
paint*

*

20 tiny gold beads

*

20 tiny pearl beads

* * * * * * *

3 Secure one wick to
one chopstick and the
two other wicks to the
other. Pour in more wax
to within 1in (2.5cm) of
the top of the bowl.

2 Place the three wicks
into the wax about 1¼in
(3cm) apart.

TIP *Once the candle has
burnt down in the
middle, you can keep
refilling it with powdered
wax and primed wick,
making it everlasting.*

1 Add vanilla scent to
the wax mixture. Pour a
little of the cream-
colored wax into the
bowl, covering an area
about 3½in (9cm) in
diameter.

*T*HIS LUSCIOUSLY
SCENTED BOWL OF WAX CAN
BE DECORATED TO SUIT
YOUR FANCY. USE PAINT,
BEADS, PEARLS, OR EVEN
TINY SHELLS FOR DELICATE
AND SUBTLE EFFECTS.

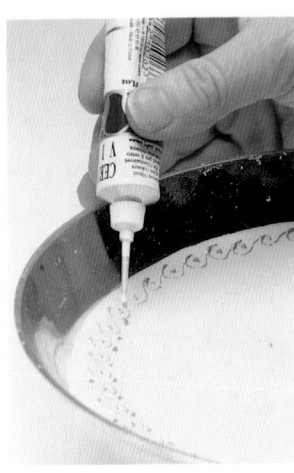

4 When the wax has set, top up the candle. You will probably have to top it up about three times before the candle has a level surface.

5 Using the silver stained-glass paint, draw a pattern of continuous "Cs" around the rim of the candle.

6 Add two rows of small dots of silver paint, one on each side of the row of "Cs." Then add a dot to the center of each "C."

7 While the paint is still wet, place gold and pearl beads alternately around the candle on the dots of paint in the middle of the "Cs."

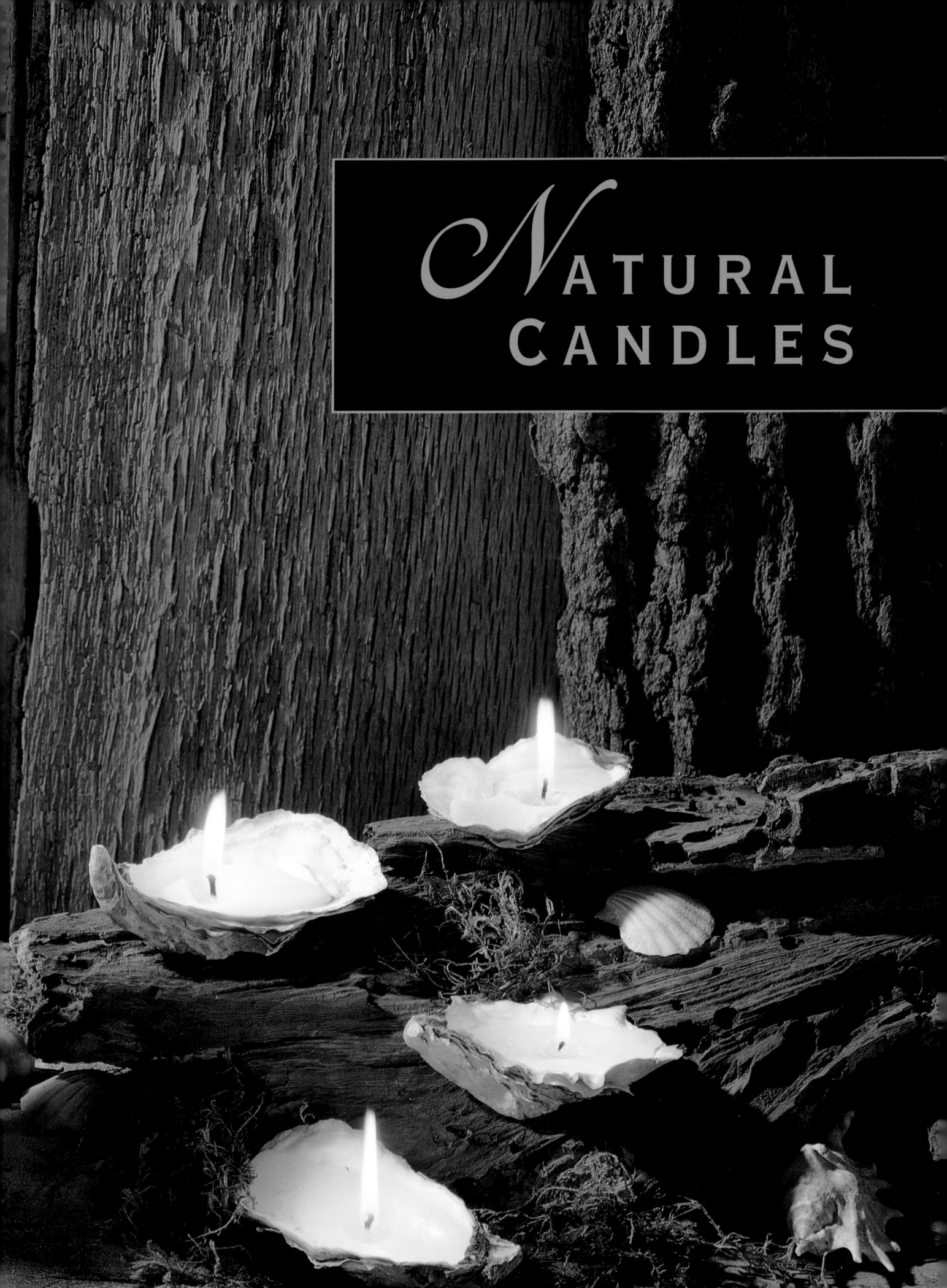

NATURAL CANDLES

BEESWAX-COVERED CANDLE

YOU WILL NEED

* * * * * * *

Beeswax sheet

Ivory-colored block candle

Ruler

Craft knife

* * * * * * *

A PLAIN, IVORY-COLORED BLOCK CANDLE COMBINED WITH THE NATURAL BEAUTY OF PLAIN BEESWAX CREATES A CLASSIC STYLE OF CANDLE WITH VERY LITTLE EFFORT.

1 Using the ruler, measure the length of the candle and mark the beeswax with the knife.

2 Using the ruler as a guide, cut out a beeswax rectangle long enough to go around the circumference of the candle with at least ⅛in (3mm) overlap.

3 Check that the rectangle is long enough before you finally trim it to size.

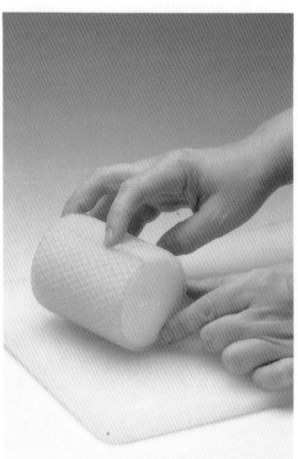

4 Wrap the beeswax sheet tightly around the candle. Starting at one end, press the two ends of the beeswax together to form a firm joint.

*T*HE NATURAL BEAUTY OF A PLAIN BEESWAX CANDLE LOOKS STUNNING IN ANY ARRANGEMENT, EVEN WHEN IT IS VERY SIMPLE.

ROLLED BEESWAX CANDLES

YOU WILL NEED

* * * * * *

2 beeswax sheets in different colors

*

Cutting board

*

Craft knife

*

Primed wick

* * * * * * *

$\mathcal{B}$EESWAX SHEETS ARE AVAILABLE IN A WONDERFUL ARRAY OF COLORS. EXPERIMENT BY COMBINING THEM IN DIFFERENT WAYS.

1 Using the craft knife, cut one sheet of beeswax diagonally across.

TIP *Beeswax is a versatile material and sticks easily to itself. Cut out letters and shapes and press them onto your finished candle. Use them to make personalized birthday or special-occasion candles.*

2 Place the second sheet on the cutting board. Place one half of the sheet you have just cut on top of it. Cut the second sheet diagonally across as before, ½in (1cm) away from the edge of the upper sheet.

3 Keeping the two sheets in the same position, turn them upside down so that the second sheet is facing you. Place the primed wick along the straight side of the candle.

4 Roll up the sheets, taking care to roll them tightly and straight. If you make a mess don't worry—you can always unroll the sheets and start again.

$\mathcal{P}$LAIN OR MULTI-COLORED, ROLLED BEESWAX CREATES AN EQUALLY MAGICAL EFFECT.

CLOCK CANDLE

YOU WILL NEED

* * * * * * *

$\mathcal{M}$ONKS AND
NUNS WERE GIVEN
THESE CANDLES SO
THAT THEY KNEW
WHEN TO GO FOR
PRAYERS OR FOR A
MEAL. ALTHOUGH THE
MARKINGS WERE
ROUGH, A 1IN
(2.5CM) WIDE
CANDLE TAKES ABOUT
AN HOUR TO BURN
DOWN 1IN (2.5CM),
SO THEY WERE
FAIRLY ACCURATE.

*Straight-sided white or ivory
candle 12 by 1in
(30 by 2.5cm)*

*

Adhesive tape

*

Ruler

*

*Sharp object, such as a small
knife, for marking candle
(see Method)*

*

*Black marker, or black
poster paint, mixed with a
few drops of dishwashing
liquid, and a paintbrush*

* * * * * * *

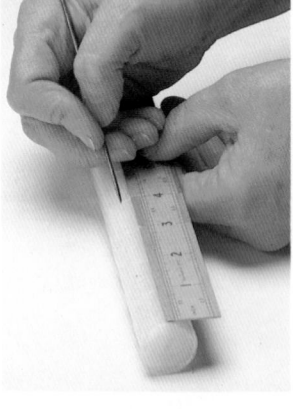

2 Place the ruler
against the candle, and
mark the candle with a
sharp object every 1in
(2.5cm) along its length.

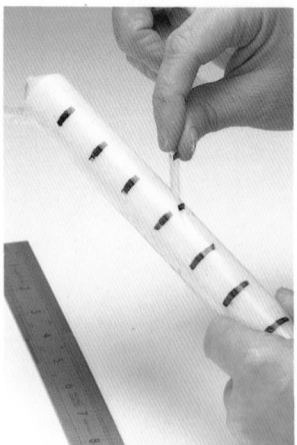

4 Remove the adhesive
tape. Insert the
completed candle into a
suitable candle holder.

$\mathcal{T}$HESE CANDLES ARE AN
ALTERNATIVE TO ELECTRIC
CLOCKS. TRY THEM NEXT
TIME THERE'S A POWER
CUT!

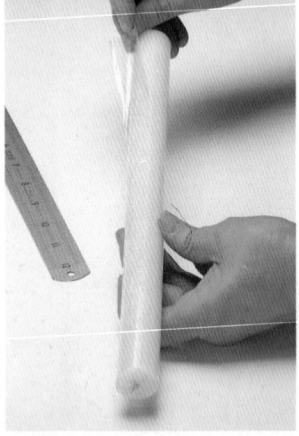

TIP *It is easier to get the
lines straight using a
marker. However, not all
felt-tips write satisfactorily
on wax, so try them first.*

1 Place two strips of
adhesive tape down the
length of the candle
leaving a gap of ½in
(1cm) between them.

3 Using the black
marker, or the poster
paint and dishwashing
liquid, make a horizontal
line on each mark.

OYSTER SHELL CANDLES

YOU WILL NEED

* * * * * * *

*T*HE NATURAL QUALITY OF BEESWAX BLENDS PERFECTLY WITH THE EXOTIC SETTING OF OYSTER SHELLS.

Oyster shells, or other large, strong shells

*

Glass or other holder to support the shell while working

*

Primed wick and wick sustainer for each shell

*

Natural beeswax (2–3oz/50–75g for each candle)

* * * * * * *

2 Melt down a small quantity of beeswax, and pour it into the shell to come about halfway up the sides.

1 Place an oyster shell on top of the glass, or other holder. Place a wick and sustainer in the center of the shell, making sure it is standing upright.

3 When the candle is almost set, check that the wick is still in the center of the candle. If not, gently pull it into the correct position. Then top up with the remaining wax.

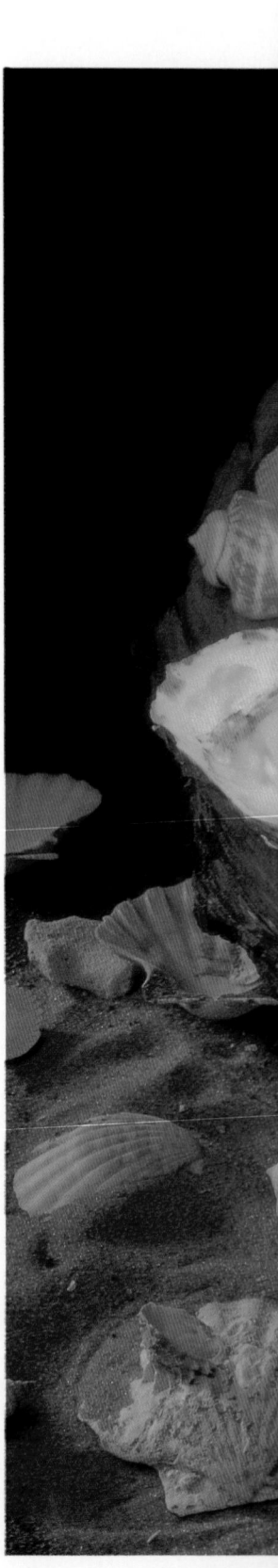

*A*NY LARGE, WIDE SHELLS ARE SUITABLE AS CANDLE HOLDERS— SCALLOP SHELLS ARE AN ALTERNATIVE.

TREASURE CANDLE

YOU WILL NEED

* * * * * * *

Candle mold

*

Container to hold candle mold

*

About 3oz (75g) dyed lilac paraffin wax

*

5oz (150g) dyed pink paraffin wax and stearin mix

*

Birthstone or other gemstone

* * * * * * *

PROVIDING YOU USE OBJECTS WHICH ARE NOT FLAMMABLE, IT IS FUN TO PLACE SURPRISE GIFTS IN A CANDLE. THESE WILL ONLY BE REVEALED WHEN THE CANDLE BURNS. THIS CANDLE USES A BLOODSTONE, WHICH IS THE BIRTH STONE FOR PISCES.

TIP *Making a layered candle will require practice. If you pour in a different-colored layer of wax when the first layer has set too hard, you will be left with a white line around the candle. If you pour it in too soon the colors will blend.*

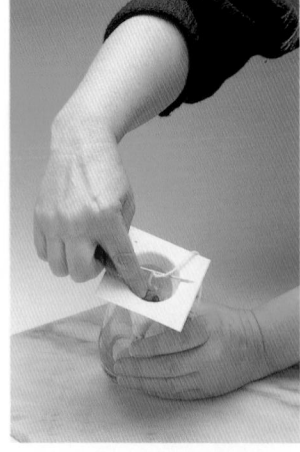

[2] Pour in a small amount of the lilac wax heated to 180°F (80°C) to about a third of the way up the mold.

[4] When the pink wax has set around the edges but is still soft in the middle, press the gemstone into the wax.

[1] Set the mold up with its wick and mold seal, and place it in the container. Prop up the container on a small object, so that it tilts at an angle of about 25°.

[3] Wait until the wax has set around the edges but is still very pliable in the middle. Now pour in a ½in (1cm) layer of pink wax, also heated to 180°F (80°C).

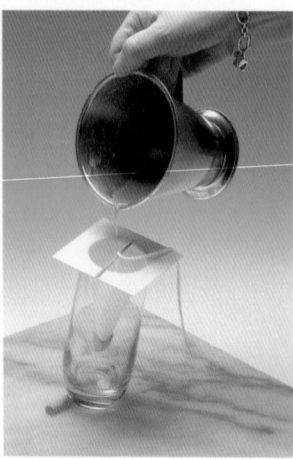

[5] Remove the prop from beneath the mold container, and stand it upright. Pour lilac wax into the mold to within ¼in (5mm) of the top. Top up in the usual way.

*S*UIT THE CANDLE TO THE SIZE AND SHAPE OF THE GIFT IT WILL CONTAIN.

ROSE QUARTZ CANDLE

YOU WILL NEED

* * * * * * *

𝒯HE VIBRATIONS EMITTED BY ROSE QUARTZ ARE SAID TO HAVE A SOOTHING AND HEALING EFFECT. PLACING THEM IN A PRETTY BOWL MAKES A CANDLE SAFE FOR THE BEDROOM.

About 8oz (225g) turquoise paraffin wax and stearin mix

*

Drop of sandalwood perfume (see page 26)

*

Pretty glass bowl

*

Wick sustainer and primed wick

*

Wicking needle or chopstick

*

Polished rose quartz stones to go around circumference of bowl

*

Scissors

* * * * * * *

3 When the wax has almost set, press the quartz stones into it all around the edge.

2 Place the wicking needle or chopstick across the center of the bowl and tie the wick to it. Pour in the wax to come about 1½in (3.8cm) below the rim.

4 Pierce the surface of the wax and then top up, being careful not to spill any wax over the stones. When set, remove the wicking needle or chopstick and cut the wick to within ½in (1cm) of the wax.

1 Add the perfume to the wax, then pour a little of the wax into the bottom of the bowl. Leave until it is almost hard, then press the wick and sustainer into the center.

$\mathcal{T}$HE COLOR OF THE
QUARTZ STONES PROVIDES
QUIET CONTRAST TO THE
WAX AND THE GLASS BOWL.

CHRISTMAS CANDLE WREATH

*H*AVING MADE THE BASIC WIRE STRUCTURE FOR THIS SEASONAL WREATH, WHICH IS VERY SIMPLE TO DO, YOU CAN ADAPT IT FOR OTHER SEASONS WITH DIFFERENT LEAVES AND FLOWERS. OR YOU COULD USE IT TO DISPLAY OTHER NATURAL FOUND OBJECTS SUCH AS SHELLS.

YOU WILL NEED

* * * * * * *

*Length of chicken wire
4 by 16in (10 by 40cm)*

*

*Circular object
(e.g. a cookie tin) to
shape the wire around*

*

Pliers

*

Wire cutters

*

*4 copper pipe end-pieces,
approximately ⅞in
(2cm) diameter*

*

*Electric drill, (to make holes
for matchsticks)*

*

8 small pine cones

*

8 matches

*

Wood glue

*

Pine needles

*

Ivy

*

*4 candles to fit copper pipe
end pieces*

* * * * * * *

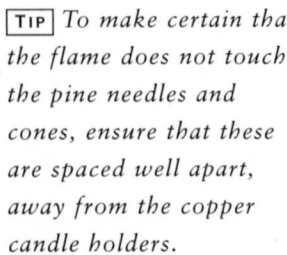 TIP *To make certain that
the flame does not touch
the pine needles and
cones, ensure that these
are spaced well apart,
away from the copper
candle holders.*

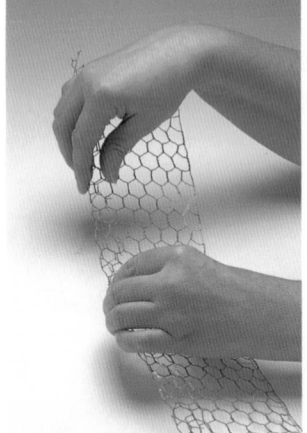

1 Roll the chicken wire so that it forms a tube.

2 Wrap it around the circular object so that it forms a perfect circle.

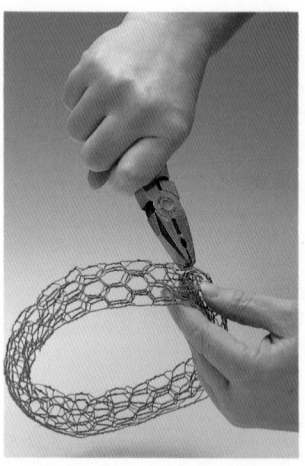

3 Use the pliers to twist the ends of the circle together.

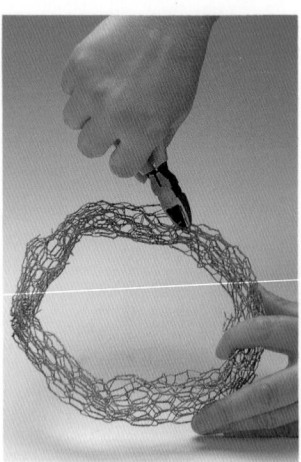

4 Use the wire cutters to cut four slits in the top of the wire. Each slit should be 1in (2.5cm) deep, and the slits should be equally spaced around the circle.

5 Press a copper pipe end-piece into each slit.

6 Drill a small hole in the base of a pine cone. Put a small dab of wood glue onto a match and insert the match into the hole. Repeat with all the pine cones. The match holds the pine cone in place in the wire.

7 Insert pine cones into the chicken wire between each piece of copper.

8 Decorate the remainder of the wreath with English ivy and sprigs of pine needles.

𝒢RESH FLOWERS CAN ALSO PRODUCE A DELIGHTFUL DISPLAY, EVEN IF IT'S ONLY TEMPORARY.

OUTDOOR CANDLES

HANGING GLASS HOLDER

YOU WILL NEED

* * * * * * *

Storage or jam jar

*

Silver glass outlining paint

*

Yellow glass paint

*

Green glass paint

*

Paintbrush

*

*24in (60cm) length of chain
(more if using a jam jar)*

*

Candle

* * * * * * *

*G*LASS PAINT CAN TURN AN ORDINARY JAR INTO A WONDERFUL FLORAL CANDLE HOLDER. A SERIES OF THESE SUSPENDED AT DIFFERENT POINTS AROUND THE GARDEN CAN CREATE A MAGICAL EFFECT ON A SUMMER NIGHT.

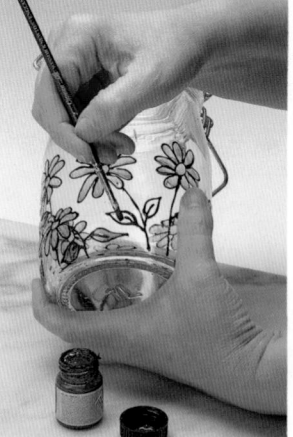

2 Fill in the flowers with yellow paint. Then fill the center of the flowers with yellow paint and the leaves with green.

3 Attach the chain to the metal clips on the storage jar to form a loop. If you are using an ordinary jam jar, wrap a length of chain around the rim first, and then attach a chain loop to this to hang the jar. Place the candle in the jar.

1 Using silver glass paint, draw an outline design of flowers and leaves on the jar.

TIP *To avoid the risk of cracking the jar when the candle is alight, the candle should fit into the jar as snugly as possible. It must not be possible to knock it against the side of the jar. The candle does not need to be secured to the base of the jar.*

*A*S THE CANDLE BURNS DOWN, THE LIGHT WILL SHINE THROUGH THE PAINT TO GIVE AN EFFECT OF ILLUMINATED FLOWERS.

CITRONELLA CANDLE

YOU WILL NEED

* * * * * * *

Flowerpot, 4in (10cm) diameter

*

About 8 oz (225 g) dyed green paraffin wax and stearin mix

*

Length of cotton or hemp sash cord. (It must not have a nylon core. If you cannot find suitable sash cord, braid three or four wicks together)

*

20 drops citronella candle scent

*

2oz (50g) undyed paraffin wax

* * * * * * *

CITRONELLA IS A POWERFUL AND REFRESHING SCENT. USED IN THIS CANDLE HOLDER IT'S A VERY EFFECTIVE OUTDOOR BUG-REPELLANT TOO — AND MORE PLEASANT AND ROMANTIC THAN CONSTANTLY SWATTING MOSQUITOES!

PLACE DIFFERENT-SIZED AND SCENTED POTS AROUND THE GARDEN FOR A FESTIVE EFFECT.

3 When the wax has hardened, pour in more wax to within 1in (2.5cm) of the top of the flowerpot. When that wax has almost hardened pull the wick to the center. Top up as often as needed to make the top level.

TIP *This candle is strictly for outdoor use. Using the sash cord as a wick makes it burn with a very large and smoky flame, which won't go out in a strong wind.*

1 Steep the sash cord in the molten undyed wax until it is thoroughly impregnated with wax.

2 Place the sash cord in the center of the flowerpot. Heat some of the green wax mixture to a temperature of no more than 170° (76°C). Mix in the citronella scent. Pour about ⅛in (3mm) wax into the flowerpot. This will both seal the hole at the bottom of the pot and fix the base of the sash cord.

OUTDOOR FLARE

A ROW OF FLARES IN A BACKYARD AT NIGHT IS A DRAMATIC SIGHT, AND ADDS TO THE PARTY SPIRIT.

YOU WILL NEED

* * * * * * *

Length of bamboo about 2½ft (75cm) long

*

Dipping can of molten clear paraffin wax

*

Length of old cotton fabric or cotton bandage

*

2in (5cm) length of wick

*

Dipping can of molten red paraffin wax

* * * * * * *

4 Dip the flare into the red wax three times, or until the cotton is no longer visible.

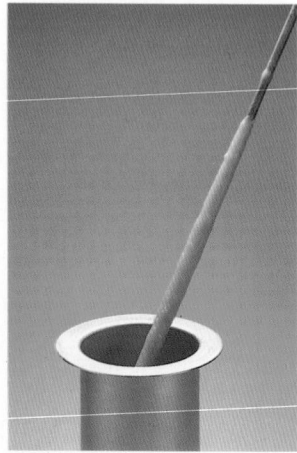

2 Wrap the fabric or cotton bandage around the wax. If you are using bandage, complete two layers. If you are using thicker cotton fabric, one layer will be enough.

TIP *Flares are a great way of using up old wax, as you never see the inside of the flare when it is burning. If you do not have a deep enough dipping can, you can pour the wax over the bamboo repeatedly to achieve the same effect.*

1 Dip the bamboo about 15 times into the can of clear wax to a depth of 12–15in (30–38cm). Stop when the layer of wax is 1in (2.5cm) thick.

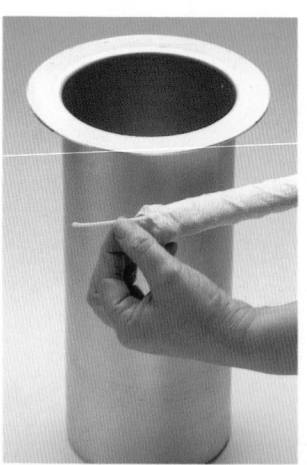

3 Push a small length of wick between the cotton and the wax.

*F*LARES CAN LAST FOR HOURS AND, PROVIDED THEY ARE SECURELY PLACED, ARE PERFECTLY SAFE.

LARGE GARDEN CANDLE

YOU WILL NEED

* * * * * *

A large round container

*

3 x 10in (25cm) lengths of 4in (10cm) wick

*

1½lbs (¾kg) green paraffin wax and stearin mix

*

Wicking needle or chop stick

* * * * * *

A SERIES OF THESE CHUNKY CANDLES CAN USEFULLY SUPPLEMENT THE OUTDOOR FLARES ELSEWHERE IN THIS BOOK.

*T*HE ADDITION OF FERN LEAVES FROM THE GARDEN ADDS SIMPLE BEAUTY AND NATURAL STYLE TO A PLAIN CANDLE.

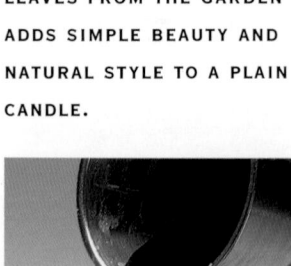

3 When the wax is half set, secure the wick by pressing it firmly down into the wax.

5 When the candle is half set, pierce its surface. Top it up until the surface of the candle remains flat. Allow to cool thoroughly.

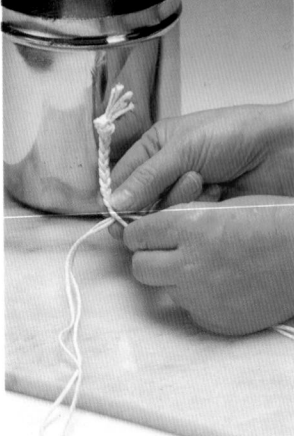

1 Tightly plait three strands of 4in (10cm) wick into one much wider wick. This ensures the candle will not blow out in windy conditions.

TIP *Make sure that you leave plenty of wick at the top. You will need it to get a good grip on the candle to pull it out.*

2 Pour in about half an inch (1cm) of wax into the bottom of the container.

4 Support the wick with the wicking needle or chopstick. Pour in the wax at a temperature of 180°F (80°C) to within half an inch (1cm) of the top of the mold.

6 Holding the candle by its wick pull it out from its mold. Trim the wick.

PAPER BAG CANDLE

YOU WILL NEED

* * * * * * *

Paper bag

*

Scissors

*

Small quantity of sand

*

Water

*

*Small block candle or
night-light candle*

* * * * * * *

CREATE A SIMPLE AND EFFECTIVE WINDPROOF CANDLE HOLDER FOR OUTDOORS. THE WET SAND KEEPS THE CANDLE AWAY FROM THE EDGES OF THE PAPER BAG.

2 Fold the bag over twice more until you have a strip with 16 layers of paper. Make a diamond-shaped cut through the folds at each side. Unfold the bag.

3 Place the bag on a flat surface and fill with about 3in (7.5cm) of dry sand.

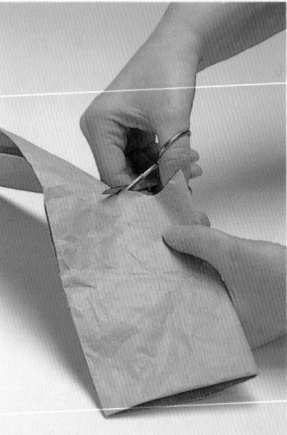

1 Fold the bag in half from top to bottom. Cut off the top, creating a wavy edge as you cut.

TIP *You can make the cuts as intricate as you like, and further decorate the bag by painting. Always place outdoor candles well away from fences or buildings.*

4 Lightly dampen the sand in the center of the bag with water.

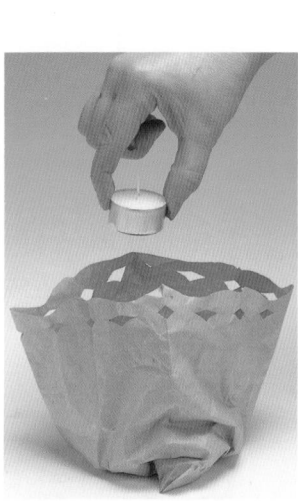

5 Insert the night-light candle or small block candle into the sand.

*T*HE CUT-OUT PATTERNS
CAN BE AS SIMPLE OR AS
COMPLEX AS YOU LIKE.
THEY TRANSFORM A PLAIN
BROWN BAG INTO
SOMETHING SPECIAL.

A POT OF CANDLES

YOU WILL NEED

* * * * * *

*T*HE LARGE WICKS OF CHURCH CANDLES MAKE THEM EXCELLENT FOR OUTDOOR USE. THEY ARE SET OFF VERY EFFECTIVELY IN THIS GARDEN OR PATIO ARRANGEMENT, BUT IT IS NOT FOR INDOOR USE.

Small bag of sand
*
Oblong terracotta pot
*
Small quantity of water
*
Two ivy plants
*
*2 church candles 6 by 2in
(15 by 5cm)*
*
*3 church candles 9 by 2in
(23 by 5cm)*
*
*2 church candles 12 by 2in
(30 by 5cm)*
*
*1 church candle 15 by 2in
(38 by 5cm)*

* * * * * *

3 Place one ivy plant on the sand at each end of the pot.

1 Pour the sand into the base of the terracotta pot and spread it evenly.

2 Dampen the sand with some water.

4 Arrange the candles in the middle of the pot. Press them well down into the sand so that they stand firmly.

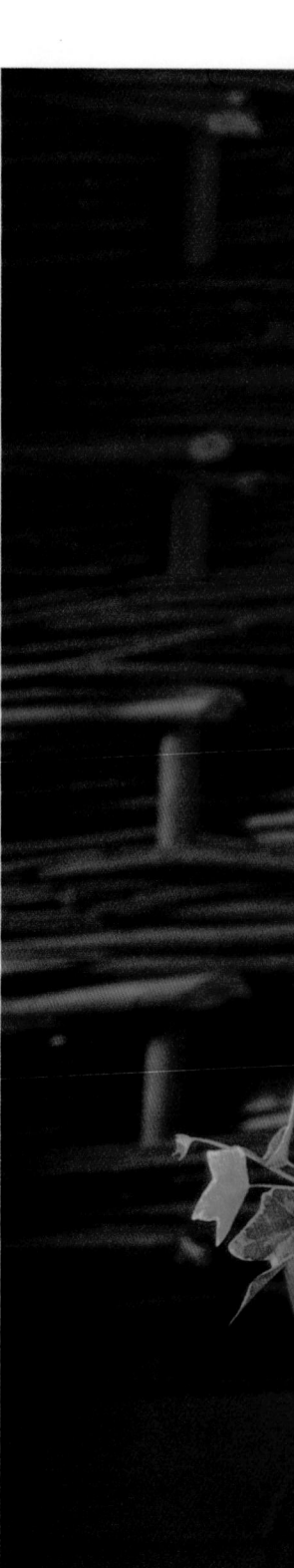

$\mathcal{T}$HE NATURAL CANDLE
COLOR IS BEAUTIFULLY
OFFSET BY THE VARIAGATED
FOLIAGE OF THE ENGLISH
IVY.

TIP *Candles burning in
groups look wonderful.
However, when they are
burnt closely together
they tend to fuse together
and drip. Arrangements
like this are for
outdoor use.*

EXOTIC CANDLES

CARVED CANDLE

*T*HIS CANDLE LOOKS COMPLICATED TO MAKE, BUT WITH THE RIGHT AMOUNT OF PRACTICE YOU SHOULD BE ABLE TO MASTER THE CARVING TECHNIQUE. IT IS A GOOD IDEA TO PRACTICE THE BASIC CUT ON MODELING CLAY BEFORE YOU ATTEMPT CUTTING A CANDLE. READ ALL THE STEPS AND MAKE SURE YOU HAVE UNDERSTOOD THEM BEFORE YOU START, AS YOU WILL HAVE TO CARVE QUITE QUICKLY TO FINISH THE CARVING BEFORE THE CANDLE SETS. ALL THE CUTS SHOULD BE MADE HOLDING THE CANDLE AT AN ANGLE OF ABOUT 30° TOWARDS YOU AND CUTTING STRAIGHT DOWN. EACH CUT SHOULD GET GRADUALLY THICKER UNTIL IT IS ⅛IN (3MM) WIDE AT ITS BASE (CUTS AT THE BASE OF THE CANDLE SHOULD BE WIDER). A CANDLE THAT HAS BEEN MOLDED OR EXTRUDED WILL BE EASIER TO WORK WITH THAN A HAND-DIPPED CANDLE, WHICH MAY FLAKE AS YOU CARVE IT.

YOU WILL NEED

* * * * * * *

10in (25cm) white candle

*

3 dipping cans or tall containers filled to a depth of at least 10½in (26cm) with 11lb (5kg) dip-and-carve wax dyed chalky white, bright yellow, and royal blue

*

Sharp knife

*

Candle varnish and brush (optional: see Method)

* * * * * * *

BEFORE YOU BEGIN
Dip the candle into the chalky white wax, then lift out. Repeat twice, then dip the candle into a bucket of cold water. Repeat the process with the yellow wax. Next dip the candle twice into the white wax again, and then into cold water. Finally, dip it three times into the blue wax—you are now ready to start carving.

1 Holding the candle steady with one hand, make a cut in the top of the candle about 2in (5cm) long and ⅛in (3mm) deep. Take the flap of wax by its tip and gently squeeze it at the top while you twist it for half a turn. Now press the top of the flap back onto the candle. Turn the candle around in your hand until you have enough space for your next cut, and repeat the process. Turn the candle again and cut and twist the last flap. You have now completed the top of the candle.

TIP *Set up a 100-watt spotlight on your work surface and carve your candle about 4in (10cm) away from it. This will help keep the wax warm and give you more time to complete the candle before it hardens.*

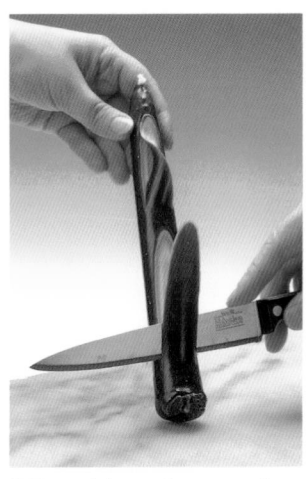

2 Holding the top of the candle with one hand, make a cut of the same depth and length as before directly below the completed cut at the top of the candle.

4 In the space left between each of the bottom and top cuts, cut a slice about 1in (2.5cm) long and ⅛in (3mm) thick. Lightly pinch the flap and bend it back to rest on the top of the bottom cut, keeping as round a shape as possible.

6 Make a circular cut around the base of the candle and remove the layers of wax below the cut to leave the original white candle exposed at the base.

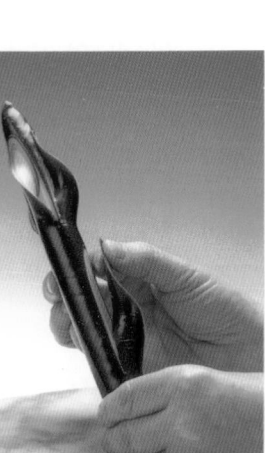

3 Squeeze and twist the flap as before and press it back onto the candle. Repeat twice, as you did when carving the top of the candle.

5 Make a second cut of the same size immediately above the previous cut. Squeeze the tip of each flap and pull it out slightly, then carefully roll it up and rest it on the flap below.

7 Lightly brush the surface of the candle with candle varnish, to give it a beautiful shine.

CARVED STAR CANDLE

YOU WILL NEED

* * * * * *

*T*HIS CANDLE USES THE SAME TECHNIQUES AS IN THE BASIC *CARVED CANDLE*. HOWEVER, BECAUSE THE CARVING IS MORE INTRICATE AND TAKES LONGER, YOU SHOULD NOT ATTEMPT IT UNTIL YOU HAVE THOROUGHLY MASTERED THE ART OF CANDLE CARVING.

Star-shaped candle about 6in (15cm) high with an even number of points

*

Meat hook and something to suspend the candle from during carving

*

5 dipping cans or tall containers, each filled to a depth of about 7in (18cm) with 11lb (5kg) of dip-and-carve wax as follows: clear undyed, and dyed chalky white, purple, bright yellow, and red

*

1 or more containers filled with hot water to keep the individual dipping cans hot

*

Candle varnish and brush (optional: see Method)

* * * * * *

1 Holding the candle by its wick, allow it to rest in the clear wax for between two and three minutes. This softens the core and makes it easier to carve. Dip the candle into the chalky white wax, and lift out. Repeat twice, then dip the candle into a bucket of cold water.

Dip three times into the red wax, twice into the white wax, and three times into the blue wax, dipping the candle in the cold water at the end of each color.

Finally, dip it once into the white wax, and you are ready to carve. Trim the excess wax from the base of the candle, and check that it stands upright. Tie extra wick onto the wick at the top and hang the candle up.

2 On one point of the star shape, make a cut about ¼in (5mm) thick and about 1in (2.5cm) from the base of the candle. Make another cut immediately above it and roll it over.

Repeat, but make the next two cuts only ⅛in (3mm) thick. You should now have three rolled flaps at the bottom of one point of the star. Repeat on the remaining points.

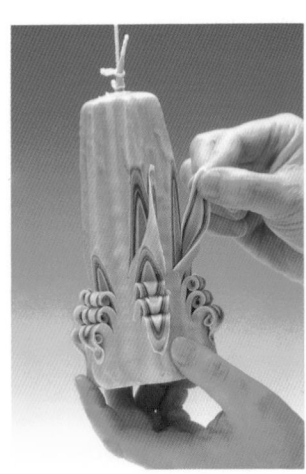

TIP *For this project, you can use either a ready-made candle or one you have made yourself.*

3 With two points of the star facing you, cut out a flap about ⅛in (3mm) deep and 2½in (6.3cm) long directly above the three bottom rolls. Gently pinch and pull the tip of the flap and give it a half twist. Repeat on the next point.

4 Holding the candle so that the two points you have just cut are facing you, cut two flaps directly above the flaps you have just made, and pull and twist them in the same way. You should now have four twisted flaps.

Take the two upper flaps you have just cut and gently pull them towards each other, pressing them down between the two points of the star facing you.

Now take the two lower flaps and bend them in the opposite direction to the two already done.

Repeat steps 3 and 4 on the remaining points of the candle.

5 Repeat what you did in step 4 above the upper flaps you have just completed, but make these flaps only about 1¼in (3.2cm) long. Twist them in opposite directions as before, and press down their tips just above the two flaps below.

You will notice that one set of flaps from the cut below is higher than the other. Bend your lower 1¼in (3.2cm) flaps onto the lower pair of flaps below. Bend the upper 1¼in (3.2cm) flaps onto the higher flaps below.

You should now have four flaps pressed together in a star shape in the gap between the two points facing you. Repeat on the remaining points of the candle.

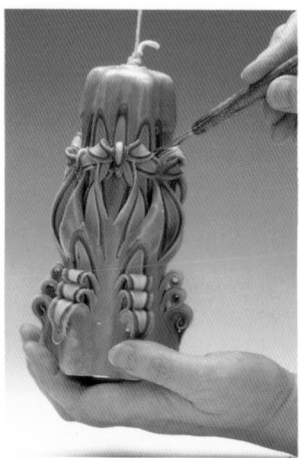

6 Hold the candle in one hand so that two points are facing you. Using a modeling tool, cut a small strip of wax at the top of the candle between the two points. Gently bend the strip of wax over the star-shaped center of the four flaps immediately below.

7 This is a close-up of the four flaps you carved in step 5 with the strip cut out in step 6 covering the center. Finally, if you want a shiny finish, you can give your candle a coat of clear varnish.

*T*HESE CARVED CANDLES ARE ONE OF THE HIGH POINTS OF THE CANDLE MAKER'S ART. THEY MAKE WONDERFUL GIFTS THAT ARE ACCEPTABLE TO EVERYONE AND LOOK BEAUTIFUL ANYWHERE.

ICICLE CANDLE

THIS DELICATE CANDLE IS MADE BY PLUNGING MOLTEN WAX INTO WATER. THE WAX, WHICH FLOATS IN WATER, SETS AS IT RISES UP TO THE SURFACE TO PRODUCE THIS EXOTIC EFFECT. EVERY CANDLE MADE THIS WAY WILL BE DIFFERENT!

YOU WILL NEED

✳ ✳ ✳ ✳ ✳ ✳ ✳

Small bowl

✳

Dishwashing liquid

✳

4–6oz (125–175g) paraffin wax and stearin mixture

✳

8in (20cm) white or ivory dinner candle

✳

Bucket filled with cold water to a depth of at least 12in (30cm)

✳ ✳ ✳ ✳ ✳ ✳ ✳

4 Holding the candle and bowl firmly, plunge them as fast as you can into the bucket of water.

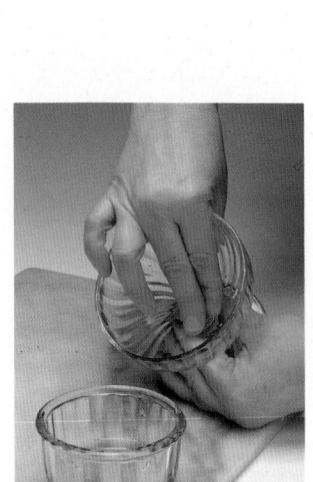

2 Pour the wax and stearin mixture into the bowl to a depth of 1in (2.5cm).

1 Using your fingers, lightly coat the bowl with dishwashing liquid.

5 Bring the candle up to the surface. Remove it from the bowl.

3 Holding the bowl with one hand, place the candle in the middle of the wax. Make sure it is standing vertically.

TIP *Sometimes the wax icicles fall sideways away from the candle. You may have a couple of seconds to adjust this, as you draw the candle up and out of the water.*

THE BEAUTIFUL AND SURPRISING SHAPES MAKE AN INTRIGUING DISPLAY AND CONVERSATION PIECE!

IVY DECORATED CANDLE

YOU WILL NEED

* * * * * * *

*T*HE SIMPLE BEAUTY OF CHURCH CANDLES IS ENHANCED BY STENCILING A NATURAL MOTIF. YOU COULD STENCIL OTHER IMAGES — MYTHICAL, RELIGIOUS OR COMIC FIGURES.

Ruler

*

Pencil

*

Sheet of laminated paper 8½ x 11 in (22 x 28cm)

*

Ivy leaf, or other small leaf

*

Craft knife

*

Church candle 12in (30cm) tall and 2in (5cm) in diameter

*

Scotch tape

*

Green poster or acrylic paint

*

Dishwashing liquid

*

Soft cloth

* * * * * * *

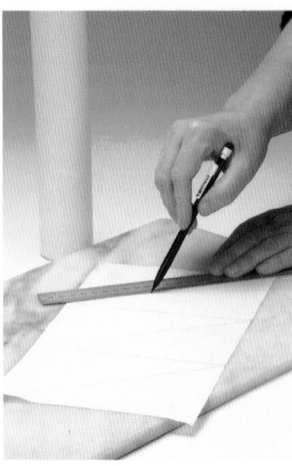

1 Using a ruler and a pencil, draw two parallel lines, one third and two thirds along the sheet of paper. Draw a diagonal line from the first parallel line to the second. Draw a second diagonal from the first to the second parallel line, and finally one from the top line to the top right-hand corner.

2 Start at the bottom left-hand corner of the sheet. Place a leaf on the diagonal line with its tip about ¼in (5mm) to the right of the line and its base laid centrally across the middle of the line. Holding the leaf in place with one hand, draw around it with the other.

Move the leaf up and repeat the process, but this time place the tip of the leaf to the left of the diagonal line. (The first leaf tip goes to the right, the second tip to the left.)

Repeat until you have leaves along all the diagonal lines.

3 Using the craft knife, cut out all the leaf shapes to make a paper stencil.

5 Prepare a mixture of 9 parts green paint to one part dishwashing liquid. Wrap the cloth around your index finger and use to apply the paint, dabbing it on from the outside to the inside of the leaf shapes.

4 Wrap the stencil around the candle and then join the ends with scotch tape.

6 Allow the paint to dry—it can take more than 24 hours—and then remove the paper.

*T*HE RELUCTANT PAINTER CAN BUY READY-MADE WAX LEAVES INSTEAD OF PAINTING. THE EFFECT WILL BE MORE WELL-DEFINED.

MOON AND STARS CANDLE

YOU WILL NEED

* * * * * * *

Ball candle

*

Small container of strongly dyed yellow paraffin wax

*

Small container of dark blue paraffin wax

*

Dishwashing liquid

*

Craft knife

* * * * * * *

*I*F YOU MAKE THE BASIC BALL CANDLE YOURSELF YOU CAN USE A THICK 1IN (2.5CM) WICK AND MAKE THIS CANDLE REFILLABLE — AND YOUR CAREFUL HANDIWORK WILL LAST FOREVER!

1 Dip the candle into the yellow wax. Lift out and dip once more.

2 Lightly smear a band of dishwashing liquid around the middle of the candle, leaving ¾in (2cm) at the top and bottom untouched.

3 Dip the candle into the blue wax. Lift out and repeat.

4 Using the craft knife, cut a small slice off the bottom of the candle so that it will stand without falling over. Using the craft knife again, draw a moon on one side of the candle.

5 Carefully lift up one end of the moon with the knife, then pull it away from the yellow base using your fingers.

*T*HIS TECHNIQUE WILL
PRODUCE DRAMATIC
EFFECTS, SO LONG AS THE
DESIGN IS KEPT STRONG
AND SIMPLE.

TIP *The moon lifts easily
off the yellow candle
because the dishwashing
liquid prevents the wax
from sticking. Use the
same principle to make
other designs—for
example, red candles
with white hearts for
Valentine's Day.*

6 To cut a star in the
blue wax, first make a
cut 1in (2.5cm) long at a
45° angle. Make a
second, similar cut to
create a V-shape, and
remove the wax. Repeat
the process once or twice
until you have a star
shape. Cut out more stars
equally spaced around
the middle of the candle.

TWISTED CANDLE

*A*DDING A TWIST TO A SIMPLE CANDLE TRANSFORMS IT INTO SOMETHING UNIQUE FOR WEDDINGS OR OTHER SPECIAL OCCASIONS.

YOU WILL NEED

* * * * * * *

10in (25cm) dinner candle

*

Deep dipping can containing about 6½lb (3kg) clear paraffin wax

*

Smooth cutting board

*

Ruler or long, smooth piece of wood or plastic

*

Large saucepan or bucket of cold water for hardening candle

* * * * * * *

2 Remove the candle from the wax; it will have become very pliable. Place it on the cutting board and lay the ruler along its length.

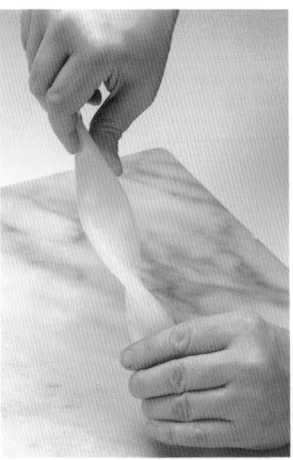

4 Holding the flattened candle by its base and its top, begin to twist it.

TIP *If the candle hardens before you have flattened it out, you can always soften it mid-stage by lowering it into the dipping can for a further minute or so.*

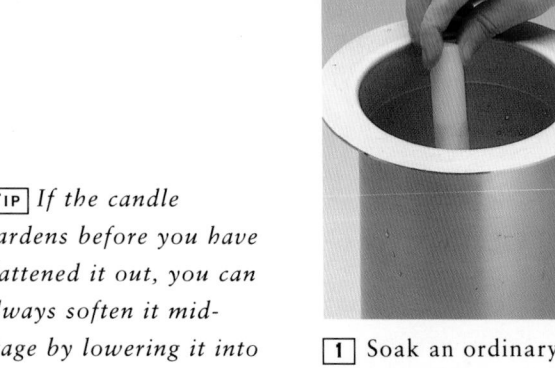

1 Soak an ordinary dinner candle in the can of clear melted wax at a temperature of about 180°F (80°C) for about 3 minutes. Do not leave it too long or it will begin to melt away.

3 Leaving the base of the candle uncovered, press the ruler down very firmly starting at the bottom end and working your way to the top, until the candle is flattened.

5 Continue twisting until the candle has been turned for a full circle.

6 Make sure that the candle is still upright and has not bent to either side, then place it to harden in the cold water.

Simply and easily made, these candles give a pleasing turn to any design feature. Their shape is displayed to best effect in simple rather than elaborate arrangements.

MARBLED CANDLE

YOU WILL NEED

* * * * * * *

THIS CLASSIC MARBLED CANDLE IS MADE USING A VERY STRAIGHTFORWARD TECHNIQUE, RESULTING IN A DRAMATIC EFFECT.

Pillar candle about 6in (15cm) tall and 2in (5cm) in diameter

*

Dipping can of molten yellow paraffin wax with 10% microcrystalline soft wax added

*

Dipping can of molten blue paraffin wax with 10% microcrystalline soft wax added

*

Container of cold water, deep enough to dip candle

*

Sharp knife

*

Dishwashing liquid

*

Round-headed hammer

*

Potato peeler

* * * * * * *

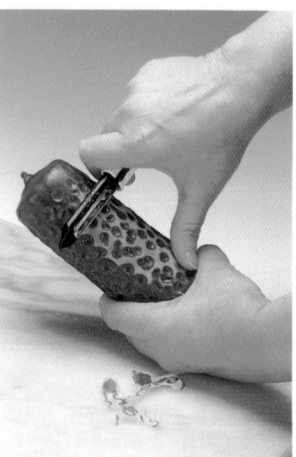

2 Trim the bottom of the candle so that it will stand straight.

1 Dip the candle four times into the yellow wax, then into the container of water. Dip twice into the blue wax, then the water. Dip twice into the yellow wax, then the water. Then dip it three times into the blue wax.

3 Rub a little dishwashing liquid onto the round-headed hammer. Then gently hammer the candle all over until the entire surface is dented.

4 Using the potato peeler, begin to carefully carve off the top layer of wax. Because the surface is uneven, you will still leave patches of blue. Take your time with the scraping. You can still scrape a bit when the wax has gone cold. The best patterns are revealed when the layers are taken off gradually.

5 Continue to scrape off the wax, to reveal a marbled pattern.

6 To smooth the candle, dip it in clear wax at a temperature of 190°F (88°C).

*T*HIS TECHNIQUE CAN PRODUCE A RANGE OF COLORS AND DECORATIVE EFFECTS VERY EASILY. THE SHAPE OF THE CANDLE WILL ALSO AFFECT THE DESIGN.

WAX BOWL WITH FLOWERS

*T*HE YELLOW FLOWERS AND DARK BLUE WAX USED IN THIS CANDLE WERE CHOSEN TO REFLECT THE COLORED SAND DESIGN OF THE BOWL.

YOU WILL NEED

* * * * * * *

6oz (180g) yellow paraffin wax and stearin mix

*

Baking tray, lightly coated with washing-up liquid

*

Chocolate mold

*

About 8oz (225g) blue paraffin wax and stearin mix

*

Bowl

*

6in (15cm) length of 1-inch (2.5cm) wick

* * * * * * *

1 Pour about ⅛in (3mm) of the yellow wax into the baking tray. While the wax is still soft, press the edge of the mold into the wax. Gently remove the shaped pieces and place on a flat surface to dry.

2 Pour 6oz (180g) blue wax into the bowl.

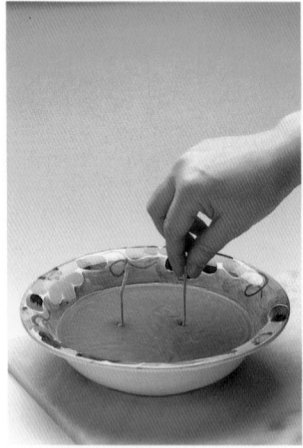

3 When the wax is almost solid, prime (see page 16) three 2in (5cm) lengths of 1in (2.5cm) wick. Take the first piece and push it down to the bottom of the still-soft wax. Carefully pull the wick back up again by about ⅛in (3mm): this prevents the base of the bowl getting too hot when the candle has burnt right down to the bottom. Repeat the process with the other two lengths of wick.

4 Gently place the yellow cut-out flowers on the surface of the candle.

5 When the candle has sunk a little in the middle, top it up very slowly, pouring extra blue wax around the yellow pieces.

*D*IFFERENT COMBINATIONS OF THE SAME COLORS WILL PRODUCE VARIED YET COMPLEMENTARY DESIGNS.

WAX FLOWER CANDLE

CREATE INTRICATE DESIGNS WITH COOKIE CUTTERS.

YOU WILL NEED

* * * * * * *

3 small baking trays

*

Dishwashing liquid

*

6oz (175g) pink paraffin wax mixed with 1½oz (38g) of microcrystalline soft wax

*

6oz (175g) pale green paraffin wax mixed with 1½oz (38g) microcrystalline soft wax

*

6oz (175g) clear paraffin wax mixed with 1½oz (38g) microcrystalline soft wax

*

1oz (25g) yellow wax

*

Flower-shaped and heart- or leaf-shaped miniature cookie cutters

*

Cake decorating pressing tool, or paintbrush

*

White block candle 3 by 6in (7.6 by 15cm)

*

Wax glue

* * * * * * *

1 Smear a little dishwashing liquid over one of the baking trays. Then pour a layer of the pink wax mixture into the tray to a depth of ⅛in (3mm). When the wax is almost hard but still pliable, cut out 20 flower shapes.

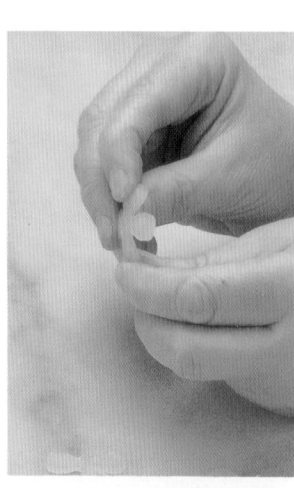

2 Take a flower shape and gently bend its petals upward.

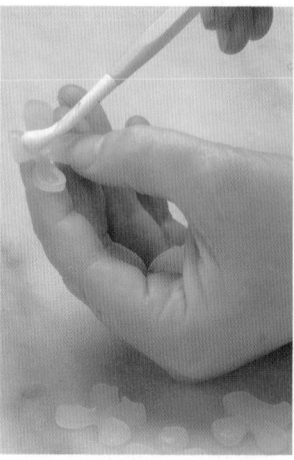

3 Using a pressing tool or paintbrush, make a small indent in each petal.

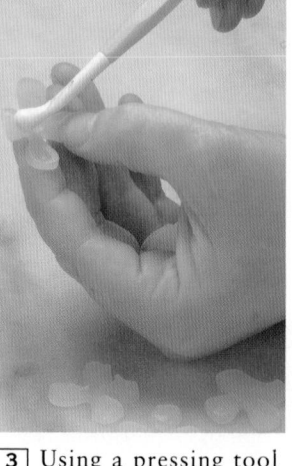

4 While the wax is still pliable, press one flower shape onto another until they join. If they have hardened too much, use a bit of wax glue to make them stick. Repeat Steps 2, 3 and 4 until you have 10 double flowers.

5 Smear a little dishwashing liquid over the second tray. Pour in a layer of the undyed wax and microcrystalline soft wax mixture to a depth of ⅛in (3mm). Again wait until the wax has almost hardened but is still pliable, and cut out 10 white flower shapes.

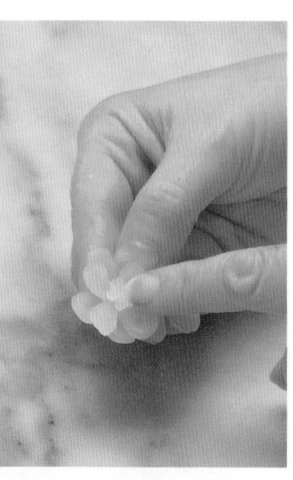

6 Press a white flower into the middle of each pink flower.

*A*LTHOUGH INTRICATE,
EACH FLOWER IS BUILT UP
IN VERY EASY STAGES, AND
THE FINAL EFFECT LOOKS
FRAGILE AND DELICATE.

7 Heat up a little
yellow wax. Use the end
of a paintbrush or the tip
of a pencil to drop
yellow wax into the
center of each flower.

8 Smear a little
dishwashing liquid over
the third tray. Pour in a
layer, ⅛in (3mm) deep, of
the green wax and
microcrystalline soft wax
mixture. Again, wait
until the wax has almost
hardened but is still
pliable and cut out the
leaf shapes.

9 Place a blob of wax
glue onto each flower
and gently press onto the
side of the candle.

10 Press two leaf shapes
onto the candle next to
each flower. Repeat Steps
9 and 10 until all the
flowers have been placed
on the candle.

NOVELTY CANDLES

MILKSHAKE CANDLE

*T*HIS DELIGHTFUL CANDLE LOOKS GOOD ENOUGH TO EAT. YOU MAY EVENTUALLY WANT TO TRY A TEQUILA SUNRISE!

YOU WILL NEED

* * * * * * *

About 5oz (150g) pink paraffin wax

*

Tall glass

*

Primed wick (see page 16) and wick sustainer

*

Cocktail stick

*

About 4oz (125g) undyed paraffin wax and stearin mix

*

2 small heatproof containers

*

Fork and spoon

*

Small quantity of red paraffin wax

*

Drinking straw or cocktail umbrella

* * * * * * *

3 Pour some undyed wax into a small container and let it cool long enough to form a skin. Using the fork, scrape off the hardened wax and beat into the molten wax in the center. Keep doing this until all the wax looks like snow. You may have to beat it quite hard to get an even consistency.

5 Heat the red wax in a container. Wait until it is half-set. Then scoop up a small amount and mold it into a round cherry shape with your hands.

1 Pour a little of the pink wax into the bottom of the glass and add the wick and wick sustainer. When the wax has almost set, center the wick with the cocktail stick. Then pour in more pink wax to within about 1½in (4cm) of the top.

2 When the wax has formed a skin about ¼in (5mm) deep, make some holes in the surface and top up.

4 Using the spoon and your finger, push the semi-molten whipped wax around the wick.

6 Place the red wax cherry you have just made into the whipped wax. Decorate the candle with a straw or cocktail umbrella.

$\mathcal{T}$HESE ATTRACTIVE CANDLES ARE IDEAL FOR A SPECIAL EFFECT AT A PARTY. A FROTHING GLASS OF BEER IS ANOTHER POSSIBILITY.

TIP *If you are giving this candle to someone else, attach a small label to it telling them to take out the umbrella or straw before they light it.*

CLOWN CANDLE

YOU WILL NEED

* * * * * *

THIS DELIGHTFUL LITTLE CANDLE IS MUCH EASIER TO MAKE THAN IT MIGHT SEEM AT FIRST SIGHT. IT MAKES AN IDEAL CHOICE OF DECORATION FOR PARTIES.

Small container of dyed bright red paraffin wax

*

Microcrystalline soft wax

*

Saucepan of boiling water for melting paraffin wax

*

Small round white candle

*

Serrated cookie cutter or chocolate mold

*

Wax glue

*

Black poster or acrylic paint

*

Dishwashing liquid

*

Fine paintbrush

* * * * * *

2 Melt a small quantity of the red wax mixed with 10% micro-crystalline soft wax in the pan of boiling water. Holding it by the wick, gently dip the white candle into the paraffin wax. Repeat three times.

3 To make the hair, press the cutter into the wax while it is still soft so that a clear impression is left. Work your way right around the candle.

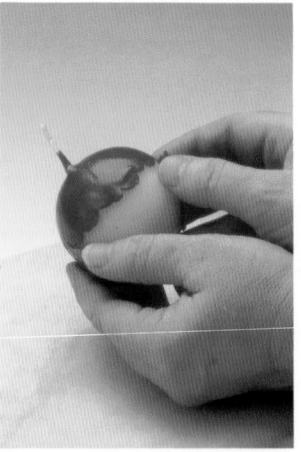

1 Holding the candle in one hand, smear dishwashing liquid gently onto the lower half of the candle.

TIP *For dipping, use a small amount of melted wax on top of water – this is a more economical use of wax. You would use considerably more wax to cover the candle without water. However, as soon as you have finished using the wax, spoon it into another container before it hardens and becomes difficult to remove.*

4 Carefully loosen the red wax below the serrated cut and peel it away. You will find that it comes off very easily as the dishwashing liquid will have prevented it from adhering to the white candle.

5 To make the nose, heat a small quantity of the red wax you used for the hair, and allow it to cool until it is solid but still malleable. Take a little of it in your hands and form it into a ball.

6 Attach the nose onto the clown's face using wax glue.

7 Squeeze out a little black poster or acrylic paint onto a dish and add about 10% dishwashing liquid. Mix the paint and dishwashing liquid together. The paint will not adhere without dishwashing liquid.

8 Carefully paint the clown's eyes and mouth.

*T*HIS CANDLE IS USEFUL FOR TRYING OUT SEVERAL DIFFERENT CANDLE-DECORATING TECHNIQUES, AS WELL AS BEING GREAT FUN TO DO.

DECORATIVE PYRAMID

YOU WILL NEED

* * * * * * *

A PYRAMID SHAPE GIVES AN UNEXPECTED SLANT TO CANDLE MAKING! WE HAVE USED A MARINE THEME TO DECORATE THIS ONE, BUT THERE ARE PLENTY OF OTHERS YOU CAN CHOOSE, A GEOMETRIC THEME, FOR EXAMPLE.

Orange paraffin wax

*

Small baking tray, smeared with dishwashing liquid

*

Cookie cutters in fancy shapes

*

Pyramid candle mold

*

About 1lb (450g) of dyed turquoise paraffin wax and stearin mix

*

Container of clear paraffin wax (optional: see Method)

* * * * * * *

3 Warm the turquoise wax to no more than 150°F (63°C), then pour it into the mold until it reaches the top. Leave to half set, then pierce the surface wax and top up in the usual way.

1 Pour a layer of orange wax about an ⅛in (3mm) thick onto the baking tray. Using the cookie cutters, cut out decorative shapes.

2 Set up the pyramid mold with the wick in the center. Smear a tiny blob of wax glue onto one side of an orange cutout. Gently press it onto one side of the mold. Do the same with the other cutouts. Repeat this process on the other three sides.

TIP *The shapes used to decorate this candle were a fish and "bubbles," but you could experiment with other shapes, such as animals, Christmas trees, Easter bunnies, hearts or flowers.*

*I*N A GROUP OR ON THEIR OWN—PYRAMID CANDLES ALWAYS GIVE STUNNING EFFECTS.

4 Remove the candle from the mold.

5 Gently dip the candle into a container of clear wax heated to about 185–190°F (85–88°C) for a few moments. This gives the candle a shine and also removes any blobs of wax glue.

MUSHROOM CANDLES

YOU WILL NEED

* * * * * * *

*Tub of green modeling
candle wax*

*

Spoon

*

*6in (15cm) length of 1in
(2.5cm) wick*

*

Wicking needle

*

*Tub of blue modeling
candle wax*

*

*Small amount of red
modeling candle wax*

* * * * * * *

$\mathcal{M}$ODELING CANDLE WAX IS A WONDERFUL MATERIAL TO USE WITH CHILDREN. WITH ADULT SUPERVISION, EVEN THE YOUNGEST CHILD CAN MAKE CANDLES WITH COMPLETE SAFETY. EXPERIMENT WITH OTHER SIMPLE SHAPES SUCH AS TREES. ADD OTHER READY-MADE SHAPES IF YOU WANT TO CREATE A FULL DISPLAY.

2 Press the base of the cylinder on a flat surface to make sure the candle stands straight. While the wax is still soft from the heat of your hands, thread the wick through the base of the candle. Leave the wick threaded through the needle.

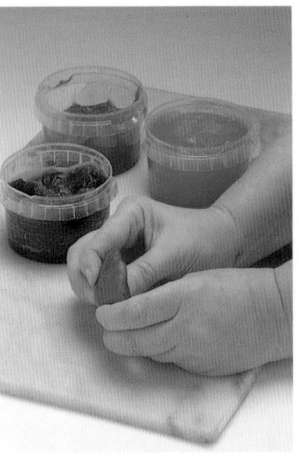

3 Spoon out about 3 sq in (7.5 sq cm) of the blue wax. Knead the wax with your hands until it is pliable. Model it into the shape of the top half of a mushroom.

1 Spoon out about 2 sq in (5 sq cm) of the green wax. Knead it with your hands until it is soft. Roll it between your hands into a cylindrical shape.

$\mathcal{T}$HESE SIMPLE SHAPES ARE VERY EASY TO CREATE AND NEED NO SCULPTING OR MODELING EXPERIENCE AT ALL.

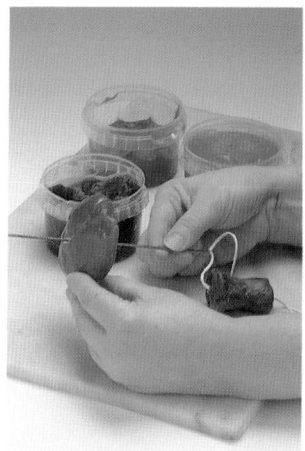 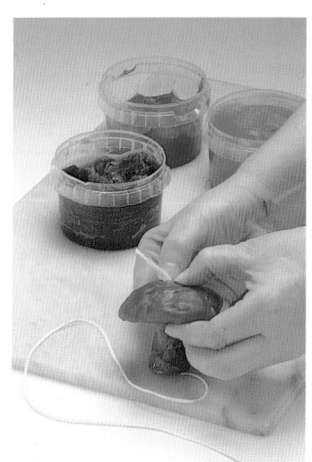 

TIP *Thread the wick through the sections of the candle as soon as you have finished modeling each one, otherwise the wax will harden and make wicking difficult.*

4 Thread the wick from the green base out through the middle of the mushroom top.

5 Pull the wick through and press the mushroom top firmly onto its base.

6 Place tiny dots of red modeling wax onto the top of the mushroom.

ASTROLOGICAL CANDLE

YOU WILL NEED

* * * * * *

𝒰SING A BLACK MOTIF ON IVORY WAX GIVES THIS CANDLE A CLASSIC BEAUTY. THERE ARE MANY DIFFERENT STAMP DESIGNS AVAILABLE, SO YOU CAN VARY THIS DESIGN TO SUIT ANY TASTE OR OCCASION.

Rubber stamp showing an astrological sign

*

Printing ink

*

Sheet of tracing paper

*

Scissors

*

Block candle 6in (15cm) tall and 2½in (6cm) wide

*

Tall dipping can filled with undyed paraffin wax

* * * * * *

2 Cut out around each print as close as possible to the edge.

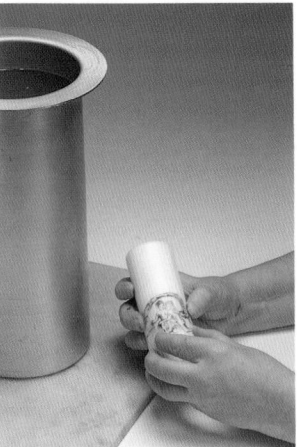

4 Take one of the prints and press it firmly into the now-soft wax of the candle. Make sure that there are no air bubbles and that the edges of the print are pressed smoothly against the candle.

1 Use the rubber stamp to print astrological designs onto the tracing paper. Vary the number according to the finished result you want to achieve, and the number of candles you are decorating.

3 Immerse the candle half-way in the dipping can of undyed paraffin wax. Keep it there for at least 30 seconds, then remove it from the can.

5 To seal the print onto the side of the candle, quickly dip the candle into the can of wax.

*G*LORIOUS GIFTS FOR
YOUR NEW-AGE FRIENDS—
AND THE TECHNIQUE
COULDN'T BE SIMPLER!

HAND CANDLE

YOU WILL NEED

✳ ✳ ✳ ✳ ✳ ✳ ✳

Rubber glove

✳

3ft (92cm) of 1in (2.5cm) wick

✳

Wicking needle

✳

Jug or jar to support the glove

✳

About 1lb (450g) red paraffin wax

✳ ✳ ✳ ✳ ✳ ✳ ✳

*T*HIS IS A WONDERFUL EXAMPLE OF HOW AN ORDINARY HOUSEHOLD OBJECT CAN BE USED TO MAKE A NOVELTY CANDLE. A GROUP OF THESE MAKES A FANTASTIC SIGHT. YOU CAN MAKE THEM OF A SINGLE COLOR, OR EXPERIMENT WITH LAYERS IN DIFFERENT COLORS.

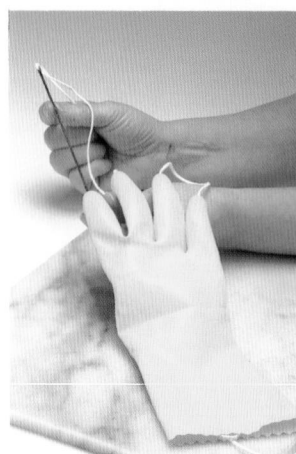

1 Thread the wick up through the little finger. Pull it until only about 2in (5cm) is showing at the bottom of the glove. Thread the wick through the top of the next finger. Pull the wick through towards the bottom of the glove only leaving a small loop between the fingers. Repeat on the other fingers and thumb.

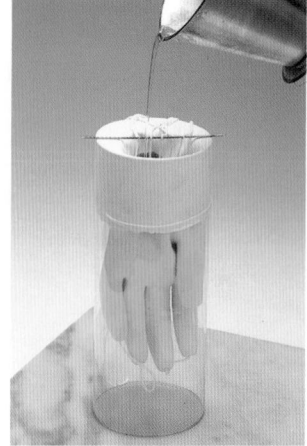

2 Pull the bottom edge of the glove over the rim of the jar or jug. Thread the wicking needle through each strand of wick, spacing the wicks ¼in (5mm) apart. Pour in the red wax up to within 2in (5cm) of the neck of the glove.

3 When the candle is half set, pierce the surface of the wax well and top up, using the remaining wax. Allow to set.

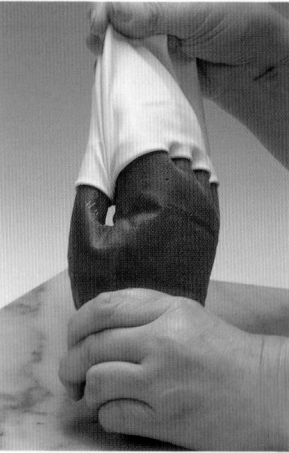

4 Cut the wicks. Pull the glove back on itself to remove the glove from the candle.

TIP *You can dip the finished hand in clear or colored wax – and fashion fingernails from wax softened with micro-crystalline soft wax.*

*G*HOULISH EFFECTS COULD ALSO BE ADDED TO THIS LAYERED CANDLE FOR HALLOWE'EN.

CANDLE HOLDERS

*T*HE BEAUTY OF CANDLES IS ENHANCED BY A THOUGHTFUL CHOICE OF HOLDER. THERE IS NOW A WIDE RANGE OF CANDLE HOLDERS AVAILABLE — IN WOOD, METAL, BRASS, TERRACOTTA, CERAMICS, AND GLASS.

INDEX

✳ For help with making the projects, thanks to William Watson and Louise Lloyd-Jones.

✳ Candle making supplies and equipment supplied by

✳ THE CANDLES SHOP (LONDON) LTD
30 THE MARKET
COVENT GARDEN
LONDON WC2E 8RE
TEL: (0) 171-836-9815
FAX: (0) 171-240-8065

✳ ANGELIC
6 NEAL STREET
LONDON WC2
TEL: (0) 171-240-2114
(candelabra p.63).